Wildlife Management

Concept, Analysis and Conservation

The Editor

Dr. Amita Saxena Professor in College of Fisheries Sciences at G.B. Pant University of Agriculture and Technology, Pantnagar, Uttarakhand India is born at Aurangabad (Maharashtra) fisheries professor, scientist took her all degrees B.Sc. (Hons), M.Sc, and Ph.D. in fish Biochemistry from Meerut University Meerut. She always ranked on top. She achieved 100%marks in marine fish keeping and tropical fish keeping diplomas from England, UK.

Dr. Amita has been awarded as an outstanding women of 20th century award (America). She gave significant contribution to Fisheries/Zoology/Aquatic Science, her researches include Fish Biochemistry, Nutrition and Fish Biotechnology. She formulated feed for air breathing and ornamental fishes/organisms/other aquatics. Her debut in pearl culture in Uttaraakhand (Hill State) and phenotypic male production of ornamental fishes research works are praise worthy and remarkable awarded by the GBPUA&T, Pantnagar. Recently she has received TWO National status Educational Excellence award. She is crowned with young scientist award, several fellowships/gold medals and Bharat Jyoti/Vijayshri, RASHTRIYA GAURAV award. She has traveled widely in India and abroad to have practical experiences and clear conception about areas, their developments, culture, traditions of the people, their needs and specially aquatic sciences of fresh water and coastal waters. She wrote 19 books, 6 six manuals several research papers, reviews, reports, thesis and delivered many invited lecture in National and International conferences. She awarded, motivated and transferred the technologies on conservation of natural resources and fisheries aquaculture to the students/common people/farmers through literature, visits, lectures, radio and TV talks. Now-a-days she is busy in improving the rural economy of poor people through fisheries & allied sectors, She is well known for her innovative ideas for teaching, research and extension for sustainable development and environment conservation. She explored the potential of aquatic resources to acquire scientific knowledge for teaching and used her research & extension skills for food, health and environment security for better livelihood for human animals and plants.

Hydrobiology- (physical and chemical properties of the water, biotopes and communities of aquatic organisms, nutrient supply of rivers, reservoirs, lakes, shallow water streams and ditches aquatic species, ecological needs of the most important aquatic taxa) how to clean these big water bodies and make artificial habitat to protect and conserve the endangered fauna and flora are her work to credit (fish farming, characteristics, fish species and farming technologies of constructed ponds as well as the biology of farmed fish species biochemistry karyotypes and molecular biology). She makes teaching creative and spontaneous and interesting and develop willingness to work collaboratively.

Induced spawning of fish, nursery fry rearing, pond farming technology. Induced spawning includes the development of the hatchery spawning technology of several farmed species. Fry rearing includes the regulation of zooplankton in ponds,. Pond farming technology includes the development of pond farming (polyculture) technology, chemical and biological inhibition of the bloom predators through teaching and training women/girls social economic development enhanced. Women's education/skill development increase the income of women and leads to growth in GDP. Educating girls leads to a number of social benefits, including many related to women's empowerment.

Preface

Certain wild species known as wild resources, are important because of their actual or potential economic value. Wild resources that provide sport in the form of hunting or fishing are known as game species. Wildlife provides people with a wide variety of direct economic benefits as sources of food, spices, favoring agents, scents, soap, cooking oil, lubricating oil, waxes, dyes, natural insecticides, paper, fuel, fibres, leathers, fur, natural rubber; medicine, fire wood *etc*.

Most of the important crops that supply 90 per cent of World's food today, were once wild plants in the tropics. Other wild species may be needed for agricultural scientists to develop new crop grains to gain higher yields and increased resistance to diseases, pests, heat and drought. Pollination by insects is essential for many food and non-food plant species. Predatory insects, parasites and microbial pathogens are increasingly used for biological control of various weeds and pests, thus keeping the loss of crops and trees under control.

About 40 per cent of drugs used throughout the world have active ingredients extracted from plants and animals. Many animals are subjected to test drugs and vaccines and to increase our understanding of human health and diseases. The nine banded Armadillo was used to study leprosy and prepare a vaccine for the disease. An estimated 10 per cent of world's marine species contain anti-cancer chemicals. Very little is known about 10-20 million undiscovered species. Less than 1 per cent earth's discovered species have been thoroughly studied to determine their possible usefulness. Loss of this biological and genetic diversity reduces our ability to respond to new problems and opportunities.

Many wild species are the source of beauty, wonder, joy and recreational pleasure for a number of people. The observation of leaves colour, in autumn, smelling aroma of wild flowers, watching an eagle soar, porpoise glide through water, Jackal call in the night, feathers of wild fowl, peacock dance, beautiful coloured spots of coccinellid beetles in jawar field, jumping style of a green coloured grasshoppers in paddy fields are pleasurable experiences that cannot be measured in terms of rupees or dollars.

The important contribution of wild species may be their role in maintaining the health and integrity of the world's ecosystems. Ecosystem services of wild plant and animals on land and in the sea include, production and maintenance of oxygen, other gases in the atmosphere and in water, filtration and detoxification of poisonous substances, moderation of earths climates, water cycles, decomposition of wastes, nutrient recycling, soil fertility, pests, diseases, photosynthesis, fossil fuel *etc.*

Wildlife resources are so rich and diverse, that a few places on the globe can be compared with it. Indian subcontinent is having sheer natural beauty and wealth and diverse species of plant and animal life. Our wildlife is spread from rain forests of Assam to barren (Thar) lands of Rajasthan and hostile mountains of high Himalayas.

This books is an unique compilation of 21 chapters, which have been contributed by leading experts who have made significant contributors to wildlife management. This book highlighted the wildlife and its conservations, which include snake and snakebites, Indian Shad, Indian Rhinoceros, Himalayan Musk Deer, Sea Turtles, Lizard, Blood Squirting Lizard, Gyps Vultures, Wild Tigers, One-horned Indian Rhinoceros, Pangolin, Indian, Bustard, Endangered Bird Conservation, Bioremediation, The Wildlife (Protection) Act, 1972 and Wildlife Conservation and its Management.

This book is valuable to students and teachers of zoology, fisheries, wildlife, professional to wildlife management, forest park rangers, animal lovers and general public to know about important wild animals.

I sincerely thanks to all contributors and those directly or indirectly have rendered their help to bring out this book. Lastly, my sincere thanks to Astral International (P) Ltd., who provides an opportunity to publish this book.

Dr. Amita Saxena

Contents

Chapter 1

Snakes and Snakes Bites

Amita Saxena

Professor,

Department of Fisheries Resource Management, College of Fisheries, GBPUAT, Pantnagar – 263 145, Uttarakhand

ABSTRACT

As the rains and warmer months approach snakes become more active.Nearly snakes are shy and will avoid contact with people if at all possible. They will only attack a human if frightened, cornered or provoked. Some of them will attack with great ferocity. If you approach a Death Adder too closely it will attack as a means of defence. THIS TOPIC IS OF TEETH GENERAL AWARENESS AND INTEREST OF THE PUBLIC IN CASE OF SNAKE BITE AND GO TO HOSPITAL.CONSULT QUALIFIED DOCTOR IMMEDIATELY.

Venomous snakes occur throughout many regions of the world and are a threat to public health, especially in the rural tropics where they are most abundant. Out of more than 3000 species of snakes in the world, some 600 are venomous and over 200 are considered to be medically important. As the rains and warmer months approach snakes become more active. Nearly snakes are shy and will avoid contact with people if at all possible.

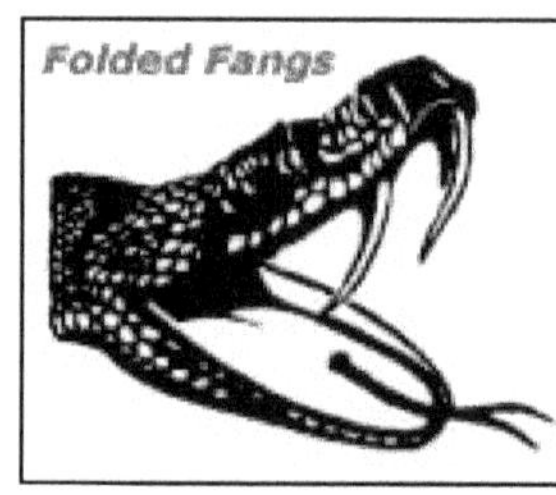

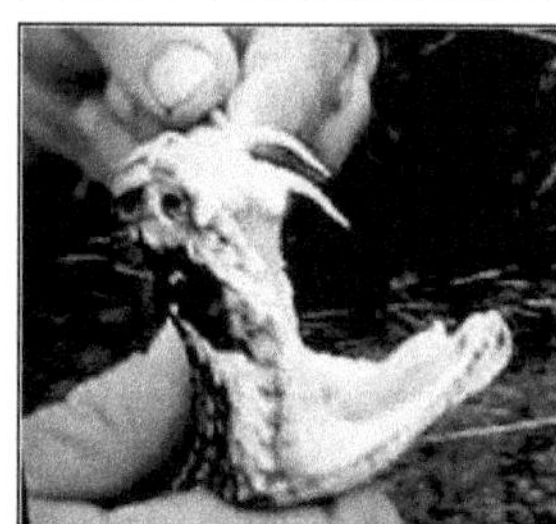

Keywords: *Snake, Bites, Treatment, Venom, Anti-venom, Poison.*

In assessing the relative risk of each species, two major categories have been considered within the WHO Guidelines on production, control and regulation of snake antivenom immunoglobulins.

The venomous snakes species listed as Category 1 are considered to cause the greatest burden of injury, disability and/or mortality attributable to snakebites in various countries, territories and other areas around the world. Species listed under this category within a country, territory or area should be considered as being of highest priority for antivenom production.

The ***proteroglypha*** have, in front of the upper jaw and preceding the ordinary teeth, permanently erect fangs.

The fixed-fang snakes (*proteroglypha*) are commonly recognized as having *neurotoxic venoms*, but there are *plenty of exceptions to this rule and it's very easy to overgeneralize. These venoms affect the nervous system,

The ***solenoglypha*** have erectile fangs; that is, fangs they can raise to an erect position. These fangs are called ***folded fangs***, making the victim unable to breathe.

The folded-fang snakes (*solenoglypha*) like rattlesnakes, cottonmouths and copperheads, ***frequently** have *hemotoxic venoms*. Hemotoxic venoms affect the circulatory system, destroying blood cells, damaging skin tissues, and causing internal hemorrhaging. It should be noted that ***not all*** folded-fang snakes have hemotoxic venoms. The majority of African pit vipers like the Gaboon Viper and numerous other species found around the world are primarily neurotoxic.

Most poisonous snakes have both neurotoxic and hemotoxic venoms and the venom dominance can vary **even within the same family** of snake. These differences within the same family of snake usually occur due to differences in habitat or geographic location and even prey species. For example, a Rattlesnake has a dominance of hemotoxic venom, while the venom of the Mexican Rattlesnakes is primarily neurotoxic. Again, this can vary greatly within any species and individual. Another example can be made with the WESTERN DIAMONDBACK RATTLESNAKE, may have a different proportion or ratio of Hemotoxic and Neurotoxic venom components from that of a Western Diamondback.

Recognizing a Venomous Bite

All poisonous snakes have 2 large fangs which are located in the upper front portion of the mouth. If the victim is bitten and the snake escapes before the identification can be made, the following signs should be noted:

One to two punctures made by the hollow fangs. Pain following within 5 to 10 minutes accompanied by swelling and discoloration around the bite area. These symptoms will progress up the victim's extremity. If the fang enters a vein or artery, these symptoms may not be present.

Coral Snake bites differ from Pit Viper bites. Their venom is neurotoxic in nature. The bite is usually not painful, little or no swelling or discoloration is present. Symptoms may be delayed for several hours but when they do occur, they progress rapidly. Symptoms include nausea, drowsiness, vomiting, marked salivation and difficulty in breathing. Paralysis is also noted in Coral Snake invenomation.

Characteristics of Pit Viper Snakes

Large fangs; nonpoisonous snakes have small teeth.

The two fangs of a poisonous snake are hollow and work like a hypodermic needle.

Pupils resemble vertical slits.

Presence of a pit. Pit vipers have a telltale pit between the eye and the mouth. The pit, a heat-sensing organ, makes it tell tale for the snake to accurately strike a warm-blooded victim, even if the snake cannot see the victim.

A triangular or arrowhead shaped head.

The rattlesnake often shakes its rattles as a warning, BUT NOT ALWAYS!

One snake that is not a pit viper snake but is poisonous is

The ***proteroglypha*** have, in front of the upper jaw and preceding the ordinary teeth, permanently erect fangs. These fangs are called ***fixed fangs***.

The ***solenoglypha*** have erectile fangs; that is, fangs they can raise to an erect position. These fangs are called ***folded fangs***.

Spitting Cobra

The **Spitting Cobra** is most renowned for their infamously remarkable ***spitting fangs*** that allow the cobra to actually spray venom accurately into the face and eyes of predators and trampling hoovestock. When the cobra wants to "Spit" or "Spray" its venom at a threat, it "Hoods Up", aims its open mouth as specialized muscles contract the **Venom Gland**, forcing the Cobra's Venom out through its fangs.

Spitting cobras are extremely accurate at distances over 10 feet. Some scientists theorize that the "Spitting Behavior" and evolution of the fang modifications were a result of the fact that Spitting cobras cohabitate with numerous antelope species. These scientists believe that the primary function of Spitting cobra fangs is to deter being trampled by antelope hooves. The Cobra is well equipped to spray its painful venom directly into the eyes of potential trampling animals from a safe distance. This theory is almost identical to the theory of the evolution of the "Rattlesnake Rattle" on rattlesnakes. Scientists speculate that the rattle's main function was to alert buffalo and other grazing North American Hoovestock to their presence in order to avoid being trampled by their feet.

Information on Spitting Cobras is plentiful as they are truly one the most remarkably adapted and unique reptiles. If you study **Pictures of Spitting Cobras**, you will notice that they come in all kinds of colors and patterns. This is due to the fact that there are many different species of Spitting Cobra. For example, there is the ***Banded Spitting Cobra*** (*Naja nigricollis nigricincta*) with contrasting bands, the small ***South African Ringhal Spitting Cobra*** that is primarily all black, the beautiful ***Red Spitting Cobra*** (*Naja mossambica pallida*), and the ***Black Necked Cobra*** (*Naja nigricollis nigricollis*), which is the most famous of all the Spitting Cobras. There are quite a few other species of Spiting Cobra.

Spitting Cobras prefer to spit or spray venom into the eyes of would be attackers, threats or predators, but are perfectly capable of delivering a deadly Snake envenomation via a bite. In fact, Spitting Cobra Bites are common in some areas of Africa and are extremely painful if not deadly. In many cases, victims of **Spitting Cobra Attacks** become permanently blind if sprayed in the eyes. Spitting Cobra venom is very toxic and extremely deadly, although it is nothing compared to the venom of the **King Cobra**, which also delivers a lot of venom with each hit. But then again, being shot by a 9mm vs. a 45mm still sucks big time, if you get my drift. So in reality, Spitting Cobra poison is the lesser of the two evils compared to King Cobra Poison.

Spitting Cobras in Captivity are much easier to care for than the King Cobra as they do not feed exclusively on snakes. Instead, Spitting Cobras feed primarily on small rodents, frogs, toads and even lizards. **Captive breeding** is achieved on a common basis with some specimens living over 20 years. Spitting Cobra eggs are laid under logs, in abandoned rodent burrows, termite nests or other suitable nooks and crannies. **Baby Spitting Cobras** are born "LOCKED AND LOADED" and pack a mean wallop if bitten. An interesting fact about captive Spitting Cobras is that the majority of them loose their desire or need to "Spray" their venom. Not that they won't bite the be-geezers out of you if given the opportunity, they will. Most cobras are very intelligent and hate wasting venom if not necessary. Perhaps captive Cobras being fed a commercial diet of thawed rodents realize their caretaker will not "Step" on them? Who knows for sure, but many Spitting Cobras seem to lose there spitting urges in captivity. Then again, there are those individual specimens that are constantly spraying venom at any passerby. These are a major pain to manage in captivity, as you always have to wipe down their enclosure.

When working within direct proximity to these snakes it is imperative to wear a FULL face protector over glasses in addition to other protective safety equipment as mere contact with Spitting Cobra venom can cause, pain, blindness, injury and even death if it gets into the eyes or an open wound. This is one reason I prefer to avoid handling these snakes. Cobras are very fast and very intelligent snakes and should ONLY be cared for in captivity by qualified experienced and State and/or Federally licensed personnel. Venomous Reptiles do not make good pets and should be appreciated from afar or in an accredited Zoological Facility. **Diamondback Rattlesnakes** are **Venomous Reptiles** and is in the family ***Crotalidae***. These impressive snakes are not only deadly, but incredibly beautiful as well. The Eastern Diamondback Rattlesnake (*Crotalus adamanteus*) unlike its counterpart the Western Diamondback (*Crotalus atrox*) at home in the arid Southwestern deserts; the Eastern Diamondback prefers high humidity and Palm Tree covered flatwoods. This magnificently colored venomous snake has even been documented swimming in salt water down.

Baby Diamondback Rattlesnakes are born deadly and measure roughly 14 inches long give or take. As they grow so do their incredibly deadly fangs! **Snake fangs** are nature's answer to the hypodermic needle, and the Diamondback Rattlesnake has been known to possess fangs in excess of over an inch! Although extremely deadly, these animals serve multiple valuable functions and deserve our protection. Rattlesnake Round Ups have decimated some species of endangered rattlesnakes, particularly the Western species. **Rattlesnake venom** varies in toxicity among the numerous species, but is primarily *Cytotoxic*, frequently referred to as *Hemotoxic Venom*. **Snake venom** is typically broken down into two categories– *Hemotoxic Snake Venom* and *Neurotoxic Venom*. Hemotoxic venom primarily attacks the blood by preventing coagulation at the same time aggressively destroying its victim's blood vessels. Neurotoxic Snake Venom acts in a completely different manner. The venom goes after the nervous system, causing massive respiratory malfunctions, heart failure and even paralysis. These symptoms are also normally accompanied by blurred vision, dry mouth, a metal taste and dizziness. One should note, that although science describes two types of snake venoms (Hemotoxic and Nerotoxic), most venomous snakes posses a combination of both in different proportions and toxicity.

While on the subject of rattlesnake venom, I feel it is important to discuss their fangs in greater detail. As mentioned earlier, snake fangs are Mother Nature's answer to the hypodermic needle. In fact, I am not certain, but I believe that the original inventors of the hypodermic needle used a snake fang as a model. Like the hypodermic needle, snake fangs have a hollow cavity running down the majority of its length. Venom from the Venom Gland then enters the snake's fang through the Venom Duct and travels down the hollow canal before it is pumped into the snakes prey or enemy from the incredibly sharp, pointed fang tip orifices. The size of these orifices or openings in the fangs has a great impact on the degree of deadliness a venomous snake possess. The size of the snakes venom gland, the venom toxicity and the size of the fangs opening are all factors that separate deadliness among the different species of poisonous snakes. For example, the Coral Snake (*Micrurus fulvius*) has one of the most potent and deadly venom of any poisonous snake drop per drop. The only problem with that is that the snake has very small venom glands, and small rear fangs with tiny fang orifices. This basically means that this snake cannot deliver large amounts of its highly toxic venom; but it also has to bite, and stay attached longer to insure a full *envenomation*. In some cases, Coral Snake envenomation has been avoided because the victim pulled the attached snake off before envenomation took place. Now compare this to the highly neurotoxic and hemotoxic venom of the Gaboon Viper (*Bitis gabonica*). This deadly African snake has the claim to fame of having not only the largest fangs of any snake (over 2.5 inches), but also the largest fang orifices as well as huge venom glands. This is one venomous snake bite to definitely avoid! Not only is the Gaboon Viper's venom extremely toxic, it also pumps in tremendous amounts of it!

Toxemia without bite *Naja nigricollis* (spitting cobra) is a species which can eject venom with considerable accuracy even from a distance of 6–12 feet. The exact range and target of this snake's venom is a matter of considerable debate among herpetologists. Most are in agreement that the venom is aimed at the victim's eyes resulting in conjunctivitis and corneal ulceration. The latter may be deep enough to cause anterior uveitis and hypopyon. There are instances on record wherein a recently killed snake and even those with severed heads have ejected venom into those handling them. This is the basis for the absolute ban on handling and extreme caution in transportation which is usually advocated for killed snakes.

Rattlesnake bites as well as other **poisonous snake bites** can be **easily avoided** if the animal is respected and only observed and appreciated from afar. Most snake envenomations occur as a result of stupidity, drunkenness and macho behavior. Like any dangerous animal, poisonous snakes need and deserve our respect. They are a vital part of the ecosystem and need protection. Not until man understands and appreciates the beauty and importance of all life, can we fully benefit from Earth's marvels and unlock her mysteries.

Poisonous Snakes King Cobras are not only extremely large, but they "Hood Out", with their face 4–5 feet in the air. When a King Cobra *strikes* at you, your heart almost stops just from the loud hiss! One's body instantly produces adrenalin upon 1st glance at a towering, hooded King Cobra locking stares with you. They almost seem to mesmerize you with their swaying bodies, trance, and eyes (almost like a beautiful feminine woman). One thing for sure is, of all the **poisonous snakes in captivity**, the Cobras are by far the most intelligent. They almost seem to have "Problem Solving" capabilities.

Speaking of **King Cobras in Captivity**, a little known interesting fact about these snakes is that they feed almost exclusively on other snakes. In fact, the 1st part of their scientific name *"Ophiophagus Hannah"*, even points out this dietary preference of consuming snakes "Ophidian". In captivity King Cobras are very difficult to care for, as "Feeder Snakes" are hard to come by in quantity and are very expensive. For these reasons, some King Cobra keepers have been successful with a feeding technique called "Odor Manipulation". With this technique, some snakes have slowly been converted to eat other easier obtained food items by rubbing dead snakes over them 1st.

King Cobra Nest is constructed by the female typically from rotten leaves and vegetation. The female King Cobra will loop her coils over loose debris and drag the material to create her nest. In most cases, a Cobra nest is nothing more then a simple pile of leaves mounded above the ground were she laid between 40 and 60 eggs. In some cases however, maybe due to experience, the female actually constructs a lower "**Egg Chamber**", with an upper area for her to lay and incubate. King Cobra eggs incubate after 65 to 80 days.

A **Baby King Cobra** is born measuring an extraordinary 18 inches in length on average and possesses a snake bite that is deadly from day 1. Baby King Cobras are highly banded with bands of white and black. These bands disappear as they mature into adults the blatant difference between adult King Cobras and baby Cobras. Looking at King Cobra Pictures is definitely the safest way to enjoy these beautiful and regal snakes. As deadly as the King Cobra bite is, these animals like all others, serve a definite function and deserve our respect and protection. Wildlife protection and conservation for all species of life, both plant and animal are deserirable.

Of all the species of pet snakes, the Milk Snake and Kingsnake are two of the most popular amongst reptile hobbyists.

Milk Snakes and some of the banded kingsnake species are often confused with the deadly Coral Snakes. As with all snakes, unless you are 110 per cent certain the snake is Non-venomous, you should keep a safe distance as extremely subtle differences in pattern or scale formation can mean life or death.

Remember, all snakes serve a major function in preserving a healthy ecosystem and should NOT be killed unless you or a loved one is in grave, and unavoidable risk of immediate injury or death. Do not kill it thinking you are a hero, doing the World a service, because you are not! Anyhow, appreciate their beauty from afar and call the appropriately licensed wildlife official. In India everything is related with god and religion. In traditional ways we worship them, protect them, this way allow these creatures to live, breed and give their contribution to the ecosystem and world.

Pet Kingsnakes and pet Milk Snakes as mentioned above are becoming extremely popular. **Captive Bred Kingsnakes** are showing up in great numbers in reptile shows around the country. These beautiful and in many cases brightly colored snakes seem to come in almost every conceivable color and pattern. Like the Corn Snake or Rat Snake, the Kingsnake is fast becoming a "Designer Snake". Breeders have become so successful in managing specific genetic traits, that they can almost create a color mutation to match your shoes. This concept of purposely creating various radical "Colour Mutations" sickens me as breeders pollute and cross contaminate the genetics of pure species.

Baby Kingsnakes are born from eggs and most species look very similar to their adult counterpart. This is also true of **Baby Milk Snakes**. This concept may seem obvious, but many baby snakes are patterned or colored nothing like the adults. For example, Baby Water Moccasins or Cottonmouths are typically born striped or banded with light colored scales. They look more like the non-venomous Banded Water Snake than their primarily solid brown or black parents.

The key is doing proper species-specific work on husbandry and its requirements well in advance of purchasing any animal. It is imperative that you have the knowledge of the reptile's dietary, habitat, lighting, temperature, caging and veterinary needs before you bring it home. Remember, the larger and more naturalistic the habitat is the healthier the animal in your care will be. Not to mention, the larger the habitat the easier it is to clean and maintain. Snakes, like tigers enjoy their room. Nothing is worse then seeing a captive snake in a tiny box covered in its own waste. If the snake is housed in a large habitat this can be avoided.Be sure to wear proper foot gear – hiking or heavy walking shoes – do not go barefooted or wear sandals to go exploring outdoors. Exercise care around fallen trees and when in heavy brush – look before you step!!

Do not place your sleeping bag(s) next to brush, tall grass, large boulders or trees where snakes are known to live and nest. Place your camp site and tents in a cleared area and use mosquito netting tucked well under the bag to provide a barrier.

Do not try to pick up snakes or handle them unless you have formal training and are able to handle them efficiently. Most experts advise against trying to handle even a freshly killed snakes as their nervous system may be still active and they can still deliver a bite

Poisonous Versus Nonpoisonous Snakes

No single characteristic distinguishes a poisonous snake from a harmless one except the presence of poison fangs and glands. Only in dead specimens can you determine the presence of these fangs and glands without danger.

Wilderness Survival and Education

Nature has provided us with beautiful scenery filled with a simply remarkable array of creatures that share this earth with us everyday. Education is the key to understanding, to preserving and to enjoying the marvels that surround us.It is important to be able to recognize various species of snakes and become familiar with their habits and habitats on wilderness journey. Nearly ALL snakes will avoid man, if possible. The danger of being bitten by a poisonous snake is very small compared to the hazards of malaria, cholera, dysentery or other diseases. There are, however, snakes known to aggressively attack man such as the King Cobra of Southeast Asia, the Bushmaster and Tropical Rattlesnake of South America and the famous Mamba of Africa but even these snakes do only occasionally. Most snakes bites results as a chance encounter very unexpectedly. Most will "get out of the way" given the chance unless aggravated intentionally. It is important to be aware of your surroundings and familiarize yourself with the snakes and animals that inhabit the area you are exploring. While snake "attacks" are rare they most certainly can be deadly.

Snakes are found in all tropical, sub-tropical and most temperate regions. Snakes use their venomous strikes to secure food and they also use these same venomous strikes for self-defense. Snake-bite accidents occur when you step on them, walk too close, agitate or threaten one – even if unintentionally.Venomous snakes occur throughout many regions of the world and are a threat to public health, especially in the rural tropics where they are most abundant. Out of more than 3000 species of snakes in the world, some 600 are venomous and over 200 are considered to be medically important

I AM WRITING IT AGAIN PREVENTION IS BETTER THAN CURE.

To Avoid Snake Bites

Don't sleep next to bush, tall grass, large boulders, or trees. They provide hiding places for snakes.

Place your sleeping bag in a clearing. Use mosquito netting tucked well under the bag. This netting should provide a good barrier.

Don't put your hands into dark places such as rock crevices, heavy brush or hollow logs, without first investigating.

Don't step over a fallen tree. Step on the log and look to see if there is a snake resting on the other side.

Don't walk through heavy bush or tall grass without looking down. Look where you are walking.

Don't pick up any snake unless you are absolutely positive it is non-venomous.

Don't pick up freshly killed snake without first severing its head. The nervous system may be still active and a partially dead snake can deliver a deadly bite.

Even if you see a venomous snake don't panic always be clam, make no sound nor movements and let the creature go its way.

Never try to kill nor hit snake with stones. In case if you do so, possibility of a jump bite or spitting of venom from snake's self-defense may prove hazardous. Spitting Cobra is very well known to spit its venom with extreme accuracy. It always target victim's eyes and its venom can cause permanent blindness

Signs and Symptoms

The bite site is usually painless. It may have classical paired fang marks, but this is not the most common picture. Often there are just a few lacerations or scratches, and sometimes these may be painless or go unnoticed. Bruising, bleeding, and local swelling may be present, but significant local tissue destruction.

The usual sequence of systemic symptom development is as follows, however it may vary dramatically:

- ☆ **<1hr:** Headache (an important symptom), irritability, photophobia, nausea, vomiting, diarrhoea, confusion; coagulation abnormalities; occasionally sudden hypotension with loss of consciousness.
- ☆ **1-3hrs:** Cranial nerve paralysis (ptosis, diplopia, dysphagia etc), abdominal pain, haemoglobinuria, hypertension, tachycardia, haemorrhage.
- ☆ **>3hrs:** Limb and respiratory muscle paralysis leading to respiratory failure, peripheral circulatory failure with pallor and cyanosis, myoglobinuria, eventually death.

Treatment of Suspected Snake bites CALL, CONSULT AND TAKE THE PATIENT TO QUALIFIED DOCTOR. DO EVERY THING AS PER MEDICAL ADVICE.

The steps for snake bite are as follows;

Some Precautions

Do not wash the bitten area.

Snake venom cannot penetrate intact skin and any residue around the site of the bite can be tested to identify the type of snake.

Stop Lymphatic Spread

A 75 mm or 100 mm bandage (preferably heavy weight crebe) is applied directly over the site of the bite about as firm as would be applied to a strained ankle.

The bandage is then continued up the full length of the limb at that same pressure and back again until the whole limb is covered. The number of bandages required will depend on the size of the limb affected.

It is considered that the bandaging should progress from the site of the bite towards the heart. The reason for this is that the progressive pressure of the bandage will tend to empty veins, thus making the limb more comfortable. It is extremely unlikely that any venom will be forced through the tissues ahead of the bandage.

Over 95 per cent of all snake bites are to the limbs (75 per cent to the legs) which allows for the effective treatment by this method. Should a bite occur on the trunk, constrictive bandaging is not recommended as it could restrict the patients breathing and breathing difficulties are usually one of the symptoms observed. You should physically apply pressure to the site of the bite with a fairly large pad and transport the patient TO HOSPITAL without delay while constantly maintaining this pressure.

Keep Immobilised

The limb should then be immobilised by the application of a splint. This is most important as any muscle contractions during movement will hasten the spread of venom by the pumping action of the muscle.

Reassurance

Reassure the patient and arrange for transport to be brought to them to avoid any unnecessary muscle movement.

If possible advise the doctor or hospital of the impending arrival and transport the patient safely and without panic to medical aid. It is most important that the driver is aware that once constrictive bandages have been correctly applied the patient should be out of immediate danger from the snake bite.

Under no circumstances should the bandage be released until the appropriate medical facilities have been prepared and the patient is under the care of a medical practitioner. Once the bandage is released it allows the trapped venom to suddenly progress unhindered and this can result in what is termed "Toxic."

Snake Bite

It is estimated that the true incidence of snake envenomation could exceed 5 million per year. About 100,000 of these develop severe sequelae. The global disparity in the epidemiological data reflects variations in health reporting accuracy as well as With the possible exception of the psychological trauma of being bitten, local changes are the earliest manifestations of snake bites. Features are noted within 6-8 minutes but may have onset up to 30 minutes. Local pain with radiation and tenderness and the development of a small reddish wheal are the first to occur. This is followed by oedema swelling and appearance of bullae – all of which can progress quite rapidly and extensively even involving the trunk Tingling and numbness over the tongue, mouth and scalp and paraesthesias around the wound occur mostly in viper bites. Local bleeding including petechial and/or purpuric rash is also seen most commonly with this family. Regional lymphadenopathy has been reported as an early and reliable changes. Generally Elapid bites result in early gangrene-usually-wet type whereas vipers cause dry gangrene of slower onset; gangrene in a limb different from the one bitten – bites were by Russell's viper. Secondary infection including tetanus and gas gangrene may also result.sign of systemic poisoning. The local area of bite may become devascularized with features of necrosis predisposing to This has relevance on the management issues. Quite often, the victim who has ventured into open fields or dense undergrowth is bitten by a species which is not immediately identifiable. In addition, the psychological reaction generated by this unexpected event impels him/her to flee: thereby further reducing the probability of confirming the snake-bite. Therefore, in a patient presenting with history suggestive onset of gangrenous the diversity of economic and ecological conditions Ophitoxaemia (Venomous snakebite).

Ophitoxaemia is the rather exotic term that characterizes the clinical spectrum of snake bite envenomation. Of the 2500-3000 species of snakes distributed worldwide, about 500 are venomous. Based on their morphological characteristics including arrangement of scales, dentition, osteology, myology, sensory organs, etc., snakes are categorized into families. The families of venomous snakes are Atractaspididae, Elapidae, Hydrophidae and Viperidae. The major families in the Indian subcontinent are: Elapidae which includes common cobra, king cobra and krait, Viperidae which includes Russell's viper, pit viper and saw-scaled viper and Hydrophidae (the sea snakes). Of the 52 poisonous species in India, majority of bites and consequent mortality is attributable to 5 species, *viz. Ophiophagus hannah* (king cobra), *Naja Naja* (common cobra), *Daboia rusellii* (Russell's viper), *Bungarus caeruleus* (krait) and *Echis carinatae* (saw-scaled viper). There are 14 venomous species in Nepal. These include pit vipers (5 species), Russell's viper, kraits (3 species), coral snake and 3 species of cobra including the king cobra.

It has been reported that in most developing countries, upto 80 per cent of individuals bitten by snakes first consult traditional practitioners before visiting a medical centre. Owing to the delay several victims die during transit to the hospital, the large majority (90 per cent) are in aged 11-50 years., predominance of male victims suggests a special risk of outdoor activity.

The high incidence of snake bite between 0400 hours to midnight corresponds well with the period of maximum outdoor activity. The incidence of snake bite shows a distinct seasonal pattern closely related to rainfall and temperature which compels the reptiles to come out of their fields, most individuals are unable to spot the snake due to tall grass and crops, fish ponds.

Most patients are unable to identify the snake species either because of ignorance or poor visibility in darkness.

Pathology

Snake venom, the most complex of all poisons is a mixture of enzymatic and non-enzymatic compounds as well as other non-toxic proteins including carbohydrates and metals. There are over 20 different enzymes including phospholipases A2, B, C, D hydrolases, phosphatases (acid as well as alkaline), proteases, esterases, acetylcholinesterase, transaminase, hyaluronidase, phosphodiesterase, nucleotidase and ATPase and nucleosidases (DNA and RNA)]. The non-enzymatic components are loosely categorized as neurotoxins and haemorrhagens. Different species have differing proportions of most if not all of the above mixtures – this is why poisonous species were formerly classified exclusively as neurotoxic, haemotoxic or myotoxic. The pathophysiologic basis for morbidity and mortality is the disruption of normal cellular functions by these enzymes and toxins. Some enzymes such as hyaluronidase disseminate venom by breaking down tissue barriers. The variation of venom composition from species to species explains the clinical diversity of ophitoxaemia. There is also considerable variation in the relative proportions of different venom constituents within a single species throughout its geographical distribution, at different seasons of the year and as a result of ageing. The various venom constituents have different modes of action. Ophitoxaemia leads to increase in the capillary permeability which may cause loss of blood and plasma volume into the extravascular space. This accumulation of fluid in the interstitial space is responsible for edema. The decrease in the intravascular volume may be severe enough to compromise circulation and lead on to shock. Snake venom also has direct cytolytic action causing local necrosis and secondary infection, a common cause of death in snake bite patients. The venom may also have direct neurotoxic action leading to paralysis and respiratory arrest, cardiotoxic effect causing cardiac arrest, myotoxic and nephrotoxic effect. Ophitoxaemia also causes alteration in the coagulation activity leading to bleeding which may be severe enough to kill the victim.The clinical manifestations of snake-bite occur in a wide spectrum with some bites resulting in minimal or no symptoms at all, while others are severe enough to result in systemic manifestations and even death

The most obvious explanation for a confirmed snake-bite but no clinical manifestations is bite by a non-poisonous species. However, it is well documented that a large number of poisonous species also often do not cause symptoms.

Systematic Manifestations

As mentioned previously, the most common and earliest symptom following snake bite (poisonous or non poisonous) is fright, particularly of rapid and unpleasant death. Owing to fright, a victim attempts 'flight' which unfortunately results in enhanced systemic absorption of venom. These emotional manifestations develop extremely rapidly (almost instantaneous) and may produce psychological shock and even death. Fear may cause also transient pallor, sweating and vomiting. The time onset of poisoning is similar in different species. Cobra produces symptoms as early as 5 minutes or as late as 10 hours after the bite. Vipers take slightly longer – the mean duration of onset being 20 minutes. However, symptoms may be delayed for several hours. Sea snake bites almost always produce myotoxic features within 2 hours so that they are reliably excluded if no symptoms are evident within this period.

Other systemic manifestations depend upon the pathophysiological changes induced by the venom of that particular species. As mentioned previously, based on the predominant constituents of venom of a particular species, snakes were loosely classified as neurotoxic (notably cobras and kraits), hemorrhagic (vipers) and myotoxic (sea snakes). However, it is now well recognized that such a strict categorization is not valid as each species can result in any kind of manifestations. Neurotoxic features are a result of selective d-tubocurarine like neuromuscular blockade which results in flaccid paralysis of muscles. Cobra venom is, however, 15-40 times more potent than tubocurarine Ptosis is the earliest neuroparalytic manifestation followed closely by opthalmoplegia. Paralysis then progresses to involve muscles of palate, jaw, tongue, larynx, neck and muscles of deglutition-but not strictly in that order. Generally muscles innervated by cranial nerves are involved earlier. However, pupils are reactive to light till terminal stages. Muscles of chest are involved relatively late with diaphragm being the most resistant. This accounts for the respiratory paralysis, which is often terminal. Reflex activity is generally not affected in ophitoxaemia and deep tendon jerks are preserved till late stages. Onset of coma is variable, however several cases of cobra bite progress to coma within 2 hours of bite. Symptoms that portend paralysis include repeated vomiting, blurred vision, paraesthesiae around the mouth, hyperacusis, headache, dizziness, vertigo and signs of autonomic hyperactivity.

Cardiotoxic features include tachycardia, hypotension and ECG changes. Cardiotoxicity occurs in about 25 per cent viperine bites and includes rate, rhythm and blood pressure fluctuations. In addition, sudden cardiac standstill may also occur owing to hyperkalemic arrest. Non-dyselectrolytemic acute myocardial infarction has also been reported. Tetanic contraction of heart following a large dose of cobra venom has been documented in vivo and in vitro. A report of non-bacterial thrombotic endocarditis following viper bite. Myalgic features are the most common presentation of bites by sea snakes. Muscle necrosis may also result in myoglobinuria.

Snake venoms cause haemostatic defects by a number of different mechanisms. Besides direct effects on the coagulation cascade, venoms also can cause qualitative and quantitative defects in platelet function. In India and Sri Lanka, Russell's viper envenomation is often associated with massive intravascular haemolysis. Haematological changes – both local as well as systemic – are some of the commonest features of snake bite poisoning. Bleeding may occur from multiple sites including gums GIT (haematemesis and melaena), urinary tract, injection sites and even as multiple petechiae. In addition cerebral haemorrhage and extradermal haematoma have also been reported. Almost every species of snake can cause renal failure. The extent of renal abnormality in them correlated well with the degree of coagulation defect; however in a majority renal defects persisted for several days after the coagulation abnormalities normalised: suggesting that multiple factors are involved in venom induced ARF.

Rarer systemic manifestations including hypopituitarism, bilateral thalamic haematoma and hysterical paralysis have also been reported.

Despite remaining well for 51/2 days. This occurred without any interference at the local site. There is also the interesting report of a zookeeper bitten on the finger following which he was administered antivenom. This prevented the development of systemic poisoning but had no effect on the extent of local complications. This individual developed compartment syndrome and spontaneous rupture of the extensor tendon of the involved finger several weeks after the bite suggesting a delayed manifestation even in the absence of systemic poisoning. ANTIVENOM VIALS ARE AVAILABLE, their doses are given to the patient as per medical advice.

Daily evaluation of patients for at least 3–4 days. Available literature suggests the use of antivenom till symptoms and signs are controlled, with some authors recommending its use as and when necessary.

Long-term Effects of Snake Bite

In most cases, swelling and oedema resolve within 2 to 3 weeks. However, they may occasionally persist up to 3 months. In exceptional circumstances, they may also be permanent. There are records, which suggest that coagulation disturbances and neurotoxicity may persist beyond 3 weeks. Necrosis of the local tissue, resultant gangrene and the consequent cosmetic defects are obvious long term effects of ophitoxaemia

Manifestations of Snake Bite not Because of Toxemia

Cases have been reported wherein the clinical manifestations of snake bite are not because of the poisoning, but due to venom hypersensitivity. This has been noted, irrespective of a history of previous bite by the same or different species. Such patients may manifest with anxiety, cutaneous sensitivity or tightness in the throat. They may also present with features of anaphylactic shock. Victims of *Bothrops* bite in individuals bitten twice developed hives and angioedema within 15 minutes of the second bite. Specific antibodies – both IgE and IgG – were detectable in their serum. The crossreactivity among the venom of *Bothrops* sp suggests that these signs are because of specific IgE antibodies against venom and must not be interpreted with toxic effects that appear late.

Factors Affecting Serverity

There are several agent, host and environmental factors that modify the clinical presentation and resultant mortality of ophitoxaemia.The number and death of the bites inflicted by the snake is a relative index of the amount of venom injected. Indirect evidence for this is also available by studying the volume of venom remaining in the glands and fangs. The condition of fangs, intact or broken, is also an indirect indicator of amount of envenomation. The species of snake which has bitten alters outcome since the amount of venom injected and the 'lethal dose' varies with species. The length of time a snake clings to its victim and the presence or absence of pathogenic organisms in its mouth are two other agent factors affecting outcome. The time of bite (day or night) and breeding habits of the snake are not related to outcome in any way. The size of snake does not appear to be related to the efficacy of envenomation since several small specimens also have lethal capacity. Among the environmental factors, the nature of first-aid and the time elapsed before administration is perhaps the single most important factor affecting outcome. The circumstances that provoked the snake to bite may also have a bearing on clinical presentation and survival of victims.

1. ***Is it actually a snake bite.*** Bite is identified by the presence of 2 puncture wounds which may vary in distance from a few millimeters to as much as 4 cms, depending on the species. The depth of the bite varies anywhere from 1-8 millimeter. In some cases, fang puncture sites are not easily

visible. They may be brought to view by Bailey's method of injecting lignocaine through a fine gauge needle and observing the sites where it oozes from. In some cases of bite, fang marks may not be visible at all. This has been attributed to a glancing strike or protection by clothing or foot wear. For the same reason, puncture wounds may even be single at times. There are instances wherein a snake has attacked repeatedly leaving multiple puncture marks. Non-poisonous snakes generally leave a row of tooth impressions, but not fangs marks.

2. ***Could it be anything else?*** Russell contends that the marks left by snakes may be so variable as to make it difficult to distinguish from bites of rats, mice, cats and even lizards. They may also be confused with insect and scorpion bites/stings. Scratches or penetration by thorns or cactus may also leave marks like those of fangs; all these may be accompanied by local changes further compounding the problem of correct diagnosis
3. ***Is it likely to be a poisonous species?*** There is no simple, reliable method to distinguish poisonous from non-poisonous species. Poisonous species generally have fangs but these may be very small in elapids and not easily visible in vipers. Tails are usually not compressed and belly scales are small in non-venomous species – all of which are opposite in poisonous species. Short of identifying the offending reptile, the only way to determine the poisonous nature of a species is to watch for features of envenomation *viz.*, local changes and/or systemic features.
4. ***Which species is involved?*** Among the commonest poisonous species in India, the cobra (nag) is easiest to identify owing to a mental picture well entrenched in most peoples minds. Technically, however it is described as having a hood bearing a single or double spectacle shaped mark on its dorsal aspect. A white band in the region where the body touches the hood is another identifying feature. The common krait (karayat) is steel blue, often shining and has a single or double white band across the back. The head is covered with large shields. In general, elapidae have relatively short, fixed front fangs; as do the Hydrophidae. Russell's viper (daboia, kander) is identified by its flat, triangular head with a white 'V' shaped mark and three rows of diamond-shaped black or brown spots along the back.

Hypofibrinogenemia may also be evident. Among the metabolic changes, hyperkalaemia and hypoxemia with respiratory acidosis, especially with neuroparalysis, urine examination could reveal haematuria, proteinuria, haemoglobinuria or myoglobinuria. In cases of ARF, all features of azotemia are also present. CSF haemorrhage has been documented in a minority of victims.ECG changes are generally non-specific and include alterations in rhythm (predominantly bradycardia) and atrioventricular block with ST segment elevation or depression. T wave inversion and QT prolongation have also been noted. In addition, cases who develop hyperkalaemia manifest typical changes of this dyselectrolytaemia.

Management under the Following Heads

1. First aid
2. Specific therapy
3. Supportive therapy

First Aid

Most physicians are in disagreement with regard to nature, duration and even necessity of first aid. Russell advises minimal wastage of time with first-aid measures which often end up doing more harm than good. Nevertheless, it is felt that reassurance and immobilization of the affected limb with prompt transfer to a medical facility are the cornerstones of first-aid care. Most experts also advocate the application of a wide tourniquet or crepe bandage over the limb to retard the absorption and spread of venom. This has relevance on the management issues. Quite often, the victim who has ventured into open fields or dense undergrowth is bitten by a species which is not immediately identifiable. In addition, the psychological reaction generated by this unexpected event impels him/her to flee: thereby further reducing the probability of confirming the snake-bite. Therefore, in a patient presenting with history suggestive of snake-bite, it is important to address the following questions. The tourniquet should be tight enough to occlude the lymphatics, but not venous drainage; though some also prefer to occlude the veins. Enough space to allow one finger between the limb and bandage is most appropriate. Should the limb become edematous, the tourniquet should be advanced proximally. Tourniquets should never be left in place too long for fear of distal avascular necrosis. In a recent report from Brazil, two cases were reported to have increased local envenoming subsequent to a tourniquet. It was formerly believed and therefore advocated that incision over the bite drains out venom. However, it has now been established from animal experiments that systemic venom absorption starts almost instantly; this form of 'therapy' is therefore being questioned. Some experts suggest that longitudinal incisions within fifteen minutes of the bite may be beneficial Suction of the local area, a staple of snake-bite management in Indian cinema, also has its advocates and detractors. While most have rejected it for its questionable efficacy, there are others who advise this method on the grounds of rapidly removing a large amount of venom. There is a patented device, the Sawyer extractor available. it is always advisable to go to hospital and take qualified medical advise.

Most authors recommend saline cleaning and sterile dressing. Some, however, advise that the wound be left open.

There is disagreement over the use of drugs as part of first-aid care. It has been suggested that NSAIDS particularly aspirin may be beneficial to relieve local pain. Russell, however, dissuades use of analgesic and in particular aspirin for fear of precipitating bleeding In Reid's experience, pain relief with placebo was as effective as NSAID. Almost all experts agree that the offending snake must not be provoked further by attempts to capture or kill it. This is for fear of provoking an already enraged reptile to strike again.

Specific Therapy – Antivenom

Antivenoms are prepared by immunizing horses with venom from poisonous snakes and extracting the serum and purifying it. Antivenoms or antivenins may be species specific (monovalent) or effective against several species (polyvalent). Monovalent antivenom is ideal but the cost and non-availability, besides the difficulty of accurately identifying the offending species – makes its use less common.

Indications for Use

There are specific indications for use of antivenom. Every bite, even if by poisonous species does not merit its use. This caution against the empirical use of antivenom is due to the risk of hypersensitivity reactions Therefore, antivenom is indicated only if serious manifestations of envenomation are evident viz coma, neurotoxicity, hypotension, shock, bleeding, DIC, acute renal failure, rhabdomyolysis and ECG changes. In the absence of these systemic manifestations, swelling involving more than half the affected limb extensive bruising or blistering and progression of the local lesions within 30-60 minutes are other indications. In a study of Elapid ophitoxaemia from India, victims with neuromuscular paralysis were administered anticholinesterase/neostigmine. Four of the patients did not receive any antivenom; all survived.

Dose

Despite widespread use of antivenom, there are virtually no clinical trials to determine the ideal dose. Conventionally 50 ml (5 vials) is infused for mild manifestations like local swelling with or without lymphadenopathy, purpura or echymosis. Thomas and Jacob have attempted to study the effect of a lower dose in a randomized controlled trial and established that, in a cohort of patients who received half the conventional dose, there is no significant difference in the time taken for clotting time to normalize. Philip also advocates using lower doses than conventionally used. Based on a study of 24 cases of demonstrated Russell's viper venom antigenemia, wherein the mean amount of monospecific antivenom correcting blood incoagualability was 165 (59.3 ml), it has been recommended that 60 ml be administered intravenously at 6 hourly intervals till blood coagulability is restored.

Administration

This matter is just for information.

Go to qualified doctor for treatment. Doses depends upon snake species. Severity of bite and person's, physiological conditions and for this doctors are best judge.

The freeze dried powder is reconstituted with 10 ml of injection water or saline or dextrose. A test dose is administered on one forearm with 0.02 ml of 1:10 solution intradermally. Similar volume of saline in the other forearm serves as control. Appearance of erythema or wheal greater than 10 mm within 30 min is taken as a positive test. In this event, desensitization is advised starting with 0.01 ml of 1:100 solution and increasing concentration gradually at intervals of 15 minutes till

1.0 ml s.c can be given by 2 hours Infusion is started at 20 ml/kg per hour initially and slowed down later.

Antivenom is administered by the intravenous route and never into fingers or toes. It has been noted to be effective in symptomatic patients even when administered up to 48 hours. Response to infusion of antivenom is often dramatic with comatosc patients sitting up and talking coherently within minutes of administration. Normalization of blood pressure is another early response. Within 15 to 30 minutes, bleeding stops though coagulation disturbances may take up to 6 hours to normalize. Neurotoxicity improves from the first 30 minutes but may require 24 to 48 hours for full recovery.

Availability

Several antivenom preparations are available internationally. In India, polyvalent antivenom prepared by C.R.I., Kasauli is effective against the 4 commonest species. Antivenom produced at the Haffkine Corporation, Parel includes more species as well. This is about 10 times as expensive as the former. The WHO has designated the Liverpool School of Tropical Medicine as the international collaborating centre for antivenom production and/or testing. Snakes do not generally attack human beings unprovoked. They are reputed to be more afraid of man than vice-versa. Nevertheless once bitten, a wide spectrum of clinical manifestations may result. The emphasis for treatment should be placed on early and adequate medical management. Overemphasis on first-aid can be dangerous because its value is debatable and too much valuable time is wasted in its administration.

SO DO NOT WORRY SNAKE BITE CAN BE TREATED. CONSULT QUALIFIED DOCTOR GO TO SPECIALIST HOSPITAL.

References

Alonso A, Scavini LM, Marino GA, Rodriguez SM. IgE antibodies against snake venoms. *J Investig Allergol Clin Immunol.* 1995, 5(1): 31–34.

Clinical trial of intramuscular anti-snake venom administration as a first aid measure in the field in the management of Russell's viper bite patients. *Southeast Asian J Trop Med Public health. 1996,* 27(3): 494–497.

Forgey WW. More on snake venom and insect venom extractors letter to the editor. *N Engl J Med.* 1993, 328: 516–517.

Gellert GA, Snake venom and insect venom extractors: an unproved therapy. *N Engl J Med* 1993, 328: 13–22.

Mirtschin PJ. Fatal cerebral haemorrhage after snake bite (letter, comment). *Med J Aust.* 1991: 155(11-12): 850–851.

Paul VK. Animal and insect bites. In: Singh M (Ed). Medical Emergencies in Children. 2nd ed. New Delhi, Sagar Publications, 1993.

Philip E. Snake bite and scorpion sting. Chapter 28 in Pediatric and Neonatal Emergency Care. Ed RN Srivatava, 1994: pp. 227–234.

Reddy KS. Essentials of Forensic Medicine an mi Printers.1980.

Reid HA, Thekaston RDG. The management of snake bite. *Bull WHO.* 1983, 63: 885–895.

Reid HA. Animal Poisons. In: Manson Bahr PEC, Apted FIC, (Eds). Manson's Tropical Diseases, 18th edition. London. Balliere-Tindall 1982, 544–546.

Reid HA. Venomous bites and stings. In: Black JA. ed. *Paediatric Emergencies.* London: Butterworths, 1979.

Russell FE. Snake venom poisoning. Philadelphia JB Lippincott Company. 1980, 325–350.

Russell FE. Snake venom poisoning. Philadelphia JB Lippincott Company. 1980, 235–285.

Thomas PP, Jacob J. Randomised trial of antivenom in snake envenomation with prolonged clotting time. *Brit Med J* 1985, 291: 177–178.

Warrel DA, Venoms, toxins and poisons of animals and plants. In: Wealtherall DJ, Ledingham JGG, Warrell DA (eds) Oxford Textbook of Medicine. 3nd ed. Vol 1, Oxford, Oxford University Press. 1996.

Win-aung, Tin-Tun, Khin-Maung-Maung, Aye-Kyaw, Hla-Pe, Tin-Nu-Swe, Saw-Naing.

Chapter 2

Declining Status and Conservation of Indian Shad *Tenulosa ilisha* (Hilsa), (Hamilton, 1822)

Seema Jain[1], Manu Varma and Nisha Rana

Department of Zoology,
R.G.P.G. College, Meerut, U.P.
[1]E-mail: seema61@gmail.com

ABSTRACT

The Indian shad Hilsa (Tenulosa ilisha) is one of the most important tropical fish of the indo pacific region. It belongs to the subfamily Alosinae of family Clupeidae and is largely an anadromous fish. Hilsa shad is anadromous, reported to ascend rivers as far as 12300km. inland for breeding and after spawning in fresh water returns to marine habitats. It inhabits coastal shelf, estuaries and freshwater rivers in Indonesia, Sumatra, Myanmmar, Bangladesh, India, Pakistan, Kuwait, Iraq and Iran. It plays an important role in the socio economic life of the northern part of Bay of Bengal. Hilsa is considered as one of the most tastiest fish due to its distinctly soft oily texture, mouth watering flavor and superb mouthfeel. The fish is locally called **'Macher Raja'** *means the king of fish.The species is widespread, but there have been serious declines in some major populations of the species due to over-exploration by commercial fisheries (especially in parts of India and Nepal), pollution in Iraq and India, and disruption of migration routes by dam development. It is still assessed as Least Concern, but further work is required to understand the species population and distribution, and the impact of fishery activities and other threats.*

Keywords: *Conservation, Decline, Ganga, Hilsa.*

The Indian Shad Hilsa (*Tenulosa ilisha*) is one of the most important tropical fish of the indo pacific region.It belongs to the subfamily Alosinae of family Clupeidae and is largely an anadromous fish. *Hilsa shad* is anadromous, reported to ascend rivers as far as 12300 km. inland for breeding (Pillay and Rose, 1963) and after spawning in fresh water returns to marine habitats. It inhabits coastal shelf, estuaries and freshwater rivers in Indonesia, Sumatra, Myanmmar, Bangladesh, India, Pakistan, Kuwait, Iraq and Iran. It plays an important role in the socio-economic life of the northern part of Bay of Bengal.

The *Tenualosa ilisha*, belongs to the subfamily Alosinae of family Clupeidae and is largely an anadromous species and is one of the most commercially important fish species of the Bay of Bengal. The species was first described and figured by Russel, 1803 from the foreshore waters of Vishakhapatnam. Then Hamilton gave the fish its first taxonomic status under the name of *Clupaudon ilisha*. It is however Regan, 1917, who created a new genus namely Hilsa and named *Hilsa ilisha*. It is capable of withstanding a wide range of salinity and travels great distances up-stream up to 1,200 km in inland water for breeding. The fish is highly priced in West Bengal, Orissa, and Tamil Nadu in India and also in Bangladesh. Hilsa shad is a filter feeder and feeds on plankton (Hora, 1938; Jones and Sujansingani, 1951). Mainly diatoms, algae and crustaceans prefer copepods in their early stages and shift their preference towards diatoms when they grew beyond 50 gm in size. It occurs in coastal waters and in the vicinity of the mouths of the river Narmada, Tapti, Cauvery, Godavari, Mahanadi, Hooghly and Brahamaputra. It ascends upstream of rivers for breeding purpose and return after spawning to the estuaries for fattening.

The largest yield of the *Hilsa* fishery comes from the deltaic region of the Gangetic system of India and Bangladesh. On account of its economic importance as well as its migratory habits, the *Hilsa* received the attention of fishery workers in India at a comparatively early date. Investigations on *Hilsa* first commenced in 1907 in the Department of Fisheries, Madras, followed by the Department of Fisheries, Bengal, Bihar and Orissa. However, very little information on the bionomics of the fish existed till 1938, when the Zoological Survey of India (ZSI) incidentally became interested on finding very small *Hilsa* in the settling tanks of the waterworks.

The studies of Hora (1938 and 1941), Hora and Nair (1940a and b) and Prashad *et al.* (1940) brought out interesting facts on the life history and bionomics of the fish. Prashad *et al.* (1940) reported the seaward migration of *Hilsa*, and pointed out that very little attention has been paid to the marine phase of the fish's life history in spite of regular commercial fishery.

The marine distribution of the *Hilsa* coincides with the Indian monsoon region. Generally, the lower Ganga river basin region is characterized by relatively large continental shelf, monsoon winds, medium to high precipitation rate and run-off, surface temperature of 20°C to 30°C, surface current changing with the change of monsoons, medium to low organic productivity, presence of sub-surface oxygen minimum layer and relatively low salinity of coastal waters (Pillay and Rao, 1962). The estuarine areas of the rivers and the brackish water lakes in the area are characterized by strong tidal action, high turbidity and heavy silting. The salinity fluctuates considerably; and in areas far from inland, the water may become entirely fresh during the monsoon months. The water level in the freshwater areas of rivers fall appreciably during the dry months; very often large areas and long stretches of the river bed are converted into chains of pools in which the fish fauna takes refuge. There is a fairly rich growth of plankton, except during the rainy season with greater abundance of zooplankton than phytoplankton (Pillay and Rao, 1962). It plays an important role in the socio-economic life of the northern part of Bay of Bengal.

The upstream migration during the main breeding season depends vitally on the commencement of the south-west monsoons. The variations in the intensity of the monsoon during the breeding season appear to cause considerable fluctuations of the fish catches in different places. Day (1873) expressed the opinion that the fish spends a part of its life in the sea not far from the shallow coastal belt. Naidu (1939) added that the general migration pattern of adult *Hilsa* take place in the Ganga from May-June and they disappear by the second week of October. Hora (1941) pointed out that among the mature *Hilsa,* which swarm into the rivers during the flood season for spawning purposes, there are a number of young individuals also and these travel far up before they become sexually mature. He stated that spawning takes place in the tidal waters and in the middle reaches of the large rivers and inferred that the floods and sexual maturity induce the fish in the sea to undertake the upward migration. Pandit and Hora (1951) summarized the extant hypothesis on the movements of *Hilsa viz.* (i) during the flood season, the adult *Hilsa* swarm up the rivers for breeding. While they probably breed in the lower of the rivers, some immature fish associated with these swarms move up and probably breed much higher up next year; (ii) the young fish fall back to the sea or estuarine areas in large numbers during October-November; the numbers get reduced by February; (iii) at about the same time, November-February, the young which had migrated to the sea during the flood season and had grown to 7"-9" form a big fishery; (iv) during March-April, the young ones known as *Jatka* enter East Bengal waters in large swarms for feeding and form an independent fishery; These swarms move up the river and fatten as they grow and mature in the middle and upper reaches, the movement being facilitated by the increase in the volume of water in the rivers due to the melting of the snow in the Himalayas during the spring hot months.

Fish passes constructed at the Farakka barrage also proved futile for the species (Malhotra and Shah, 1979). Besides, the obstructions of rivers by the construction of weirs and anicuts, silting also appears to affect the *Hilsa* migration. The occurrence of two runs of *Hilsa* in the main river system, one during the south west monsoon and the other during late winter has been established. The late winter run is of a smaller magnitude while the estuarine stocks migrate up the river for spawning and return to the river mouth and adjacent foreshore areas. Temperature, current, velocity, and volume of discharge are probably the significant directive factors to which *Hilsa* responds in its movements from the sea to the estuary (Pantulu *et al.,* 1966; Gopalakrishnan, 1973). The general assumption is that the maturing *Hilsa* ascend the river till they reach the spawning grounds and after spawning they migrate downwards.

Hilsa is known to be a fast swimmer (Southwell and Prasad, 1918). Tagging experiments have shown that the fish is capable of covering as much as 70.8 km in one day. *Hilsa* is generally found to move on the sea surface whereas in the river they move at a depth of 14-18 m (Mojumdar, 1939); though, on a cool or drizzly day they may rise to within 2 m from the surface. During their migration upstream the fishes congregate, but do not form dense shoals; however during winter months, they are found to form very large shoals. *Hilsa* assemble in large numbers below dams or other obstructions to their upstream migrations (Pillay and Rao, 1962). It was also noticed that the *Hilsa* move near the bottom of rivers and rise to the surface when they meet obstructions such as dams or anicuts (Southwell and Prasad, 1918). During, the breeding season *Hilsa* ascends the rivers and after spawning, returns to the original habitat where they remain till the next breeding season. It has been observed that the males move in the surface zone and the females in deeper areas (Pillay, 1958). Eggs, larvae and juveniles of the *Hilsa* are found in the upper reaches of the river during the south-west monsoons (July-September) and again during the second half of winter (January-February). From October till end of November and again during March, juveniles and young ones (up to 8 cm length) are found in the lower stretches and estuarine areas in lower Ganga river basin. Young fish of about 15–22 cm length occur along the foreshore areas during the winter months (December-February). The eggs occur in the sub-surface zones, while the juveniles appear to inhabit the subsurface waters, but later age groups move in deeper zones of water as well.

Decline in *Hilsa* Fishery

Loss of habitat, directly and indirectly through fishing or other processes, poses major threat to the continued existence of many marine species (Roberts and Hawkins, 1999; Rodwell *et al.*, 2003), particularly those that are already endangered (Wilcove *et al.*, 1998). Undeniably, mistreatment is alleged to have caused 55 per cent of marine extinctions, while habitat degradation explains a further 37 per cent (Dulvy *et al.*, 2003), emphasizing the importance of these two processes for both conservation, sustainability and subsistence. Some studies have concluded that habitat is more important than fishing (Grigg, 1994), whilst others have found that the effects of fishing are still prevalent when accounting for spatial variation in habitat (Chapman and Kramer, 1999). The relative consequence of these two processes is likely to depend on the extent of fishing pressure compared to variations in habitat (Russ *et al.*, 2005) and will vary depending on species vulnerability to fishing versus habitat dilapidation (Wilson *et al.*, 2008).

The river Ganga has greatly influenced human habitations along its littoral zone, supporting the livelihoods of many people, but suffered from unregulated fishing, environment degradation, water abstraction and encroachment. The amplified fishing pressure due to higher claim for *Hilsa* fish, followed by indiscriminate fishing methods increased the fishing effort leading to over exploitation, which gradually led to a drop in the catch per unit effort. With the decreasing natural stocks, the fishers had to augment fishing effort for whatever species or size of fish were available to support their livelihoods. Interventions like regulations, wise use, or increased awareness may not yield desired results, and to reverse the trend, as fisher's livelihoods are affected. Although the ecology, fish species composition and landing trends are studied in Ganga, there is also acute paucity of sound empirical information on the fish population, maltreatment levels and sustainable yields from the river to implement effectual resource management plans. It has been reported by several workers that variability of the monsoons cause considerable fluctuations in the *Hilsa* catches. The intensity of monsoon, its arrival, and the consequent flooding of the rivers along with high turbidity of the waters have been included as causal factors for *Hilsa* fishery decline. Hora and Nair (1940a) and Hora (1941) have recorded that besides the annual fluctuations, there is a five-year cycle in the *Hilsa* fishery. This view was supported by Biswas (1954) on the basis of the trend in prices. Hora and Nair have also observed that the long range fluctuations are due to largepopulations attaining maturity at particular periods. They have inferred that a majority of the fish attain maturity when they are five years old and hence every five years bulk of the stock on becoming mature swarms up the river and provides bumper catches. Dunn (1982) has also considered water level fluctuations and turbidity values, expressed as index of sediment load compared with water flow, as causal factors in the fluctuations in the *Hilsa* fishery.

In the recent years there has been a declining trend in the catches of *Hilsa* along the upstream and downstream (δ, Talsari and Frazerganj) of the Hooghly estuarine region. Being a migratory fish, the adults that swarm at the lower estuaries and mature adults that make long journeys into the rivers are subject to noticeable fishing stress. Similarly, the fry that spend their life for a season in the river and juveniles that make long trip into the sea through the river and estuary are also subjected to extreme fishing stress. It is also a matter of concern that huge quantities of pre-adults are caught in the inshore areas by *jangal* and *kachal* fishing gears even before the fish attains size at first maturity (370 mm).

Fishing on the migrating ripe *Hilsa* is also a matter of concern, causing failure of *Hilsa* fishery. The small mesh size of *jangal* and *kachal* and fishing in the river during September and February when large-scale movement of ripe fish is expected in the river, is a major cause of *Hilsa* decline. The overall rise in global temperatures has also affected the *Hilsa* migration.

In India, the *Hilsa* fishery is managed by the state, rather than by the central government. This is likely to mean that there is less capacity and resources to actively implement management measures. There is currently no control on fishing effort, that small size mesh nets are widely used to catch *jatka* and similar-sized juveniles of many species. Other scientist also undertook an assessment of the "total catchable potential" (TCP) of the Hooghly river system and some of the main species, including *Hilsa*. They estimated that the TCP for *Hilsa* was 3507.6 tons and this has already been exceeded. Thus, the limited studies on *Hilsa* suggest that *Hilsa* are almost certainly overexploited in India. Thus, if India were to use management measures to improve sustainability, then major spawning and nursery areas need to be identified and mapped. Researches and investigators have also identified pollution and poor environmental flows as serious problems that are affecting *Hilsa* and other riverine and estuarine fishes in West Bengal.

Hilsa was the main fishery in comparison to major carps during the period between 1963-1971. But, with the commissioning of Farakka barrage, the fishery declined sharply between 1972-1980. Ghosh (1976) has mentioned that the production of the *Hilsa* fishery above the Farakka barrage has dropped from 116.1 kg/km^2 pre-construction to less than 1 kg/km^2 post construction. Catches of *Hilsa* dropped to just 1.01 tons at the same time the miscellaneous fishery recorded an increase. Total fish catch from this centre registered improvement, but shift in species composition was a direct result of hydrological changes caused by barrage. The fishery declined both in quantity and quality in time scale due to various anthropogenic factors. It has been reported by Gupta and Tyagi (1992) that the fishery is being harvested at much higher effort than the optimum fishing effort. Therefore, urgent steps are required to reduce the fishing pressure to achieve the goal of sustainable fishery from the river system. *Hilsa* fishery suffers badly. The per unit yield of major carps dropped to 46.0 kg/km from 143.5 kg/km recorded during the 1960s while *Hilsa* yield was only 7.1 kg/km. With the construction of Farakka barrage, the fishery scenario at Lalgola center about 45 km. below Farakka, showed a major change in stock structure. Prior to Farakka, the *Hilsa* used to be the main fishery (92.02 per cent). With the commissioning of the barrage, *Hilsa* contribution came down to merely 16.8 per cent and the niche was replaced by other species. In recent years, between 1998-99 and 2002-2003, the average annual catch of *Hilsa* has been estimated at 10382.9 tons with an impressive increase of 63.3 per cent from the corresponding five years (6279.6 tons). The average annual landing showed a sharp increase in the post barrage period. It was 1457.1 tons (15.3 per cent) prior to 1975 and increased to 7352.9 tons (13.2 per cent) during 1994-95 to 1999-2000. This unusual increase in the catch of *Hilsa* may be also due to the barrage construction, and due to migration of the *Hilsa* to areas above the barrage, resulting in severe depletion of *Hilsa* fisheries in the middle stretches of river Ganga. *Hilsa* juveniles (fry and fingerlings) constitute a substantial part of *Hilsa* catch from the upper freshwater stretches of the estuary.

Indiscriminate exploitation of the young ones of *Hilsa* through small mesh nets took a heavy toll on the *Hilsa* juveniles, on the onset of downward migration of the young ones. Mitra *et al.* (2001) estimated *Hilsa* juvenile catch between 50.9–63.3 tons (57.5 tons) from Hooghly estuary, during the period 1994–1995 to 1999–2000, numerically when estimated it results to 13.1 million of young fish (weight range 2.2–27 g). In addition to this, wanton killing of *Hilsa* juvenile has been rampant in the upper stretch of the estuary, especially during November to May due to development of small mesh net.

Low water discharge from the river Ganga at the Farraka barrage and associated heavy siltation, indiscriminate exploitation of juveniles (*jatka*), disruption of their migration routes, loss of spawning, feeding and nursery grounds and increased fishing pressure have all contributed to a decline in the catch per unit effort in both the marine and river *Hilsa* fishery. The radical decrease of catches of both mechanized and non-mechanized boats indicate the excess of fishing effort, which could lead to overexploitation and vulnerability of the fishery. The declining trend of catch per unit effort of *Hilsa* fishing is threatening the livelihoods of about 464 thousand *Hilsa* fishermen. Fish stocks are renewable and a pragmatic approach is essential to maximize the sustainable benefits they can generate. It must be ensured that the resources are protected from the irreversible damage and managed on a sustainable basis. Dams and barrages constructed across the river to supply irrigation water, flood protection and hydropower not only prevent migration, change migration routes, and alter spawning and nursery grounds, but also concentrate the shad population in certain areas, thereby subjecting them to overfishing.

Conclusions

Thus, it can be concluded that the *Hilsa* decline thoroughly indicates habitat destruction and degradation in the lower Ganga river basin and at the same time it projects the rapid alteration in the ecology of the section concerned. The existing situation of the *Hilsa* fishery suggests that proper assessment is necessary and there is a need to find a way to stimulate the recovery of the fishery and make it sustainable, while maximizing economic benefits.

Construction of barrages on the rivers have significantly checked it's upwards migration. For example construction of Farraka barrage on ganga at Farraka in 1975 has obstructed upward migration of the fish.Thus declining its production on important upstream landing centres (Lalgola, Bhagalpur, Allahabad).

There is a compelling need for domestication and culture of the species for conservation and rehabilitation of its declining fishery.

References

Day, F. 1 873a Report on the Freshwater Fish and Fisheries of India and Burma. Calcutta: pp. 22, 23, 35–36.

Day, F. 1 873b Report on the Sea Fish and Fisher/es of India and Burma. Calcutta: pp. 24, 25 and App. pp. i, xxxi. cxxi & cxxii.

Dulvy, N.K., Sadovy, Y. and Reynolds, J.D. 2003– Extinction vulnerability in marine Populations. *Fish*, 4:25–64

Dunn, I. C. 1982a Hilsa resources development and management in Bangladesh. Fl:DP/BGD/72/016, Field document 2, FAO, Rome : 47 pp.

Dunn, I. C. 1 982b The H/Isa fishery of Bangladesh, 1982: An investigation of its present status with an evaluation of current data. Fl:DP/BGD/81/034. Field document 2, FAO, Rome: 71 pp.

Ghosh, K. K. 1976 Inland fishery resources of India, their estimation and utilization. Cent.Inland Fish. Res. Inst., Barrackpore.

Gopalakrishnan, V. 1969 Observations on the present status of hilsa fisheries in the Hooghly–Matlah estuarine system (West Bengal, India). *Proc. Indian Sci. Cong.*, 56(3): 539–540.

Gopalakrishnan, V. 1973 Fishery resources of the Hooghly–Matlah estuarine system and its relation to fisheries of the Bay of Bengal. Proc. Symp. Living Resources of the Seas around India, Spl. Pub. CMFRI: 373–386.

Hora, S. 1. 1938 A preliminary note on the spawning grounds and bionomics of the so–called Indian shad, H/Isa i/isha (Ham.), in the river Ganges. *Rec. Indian Mus.*, 40(2): 147–1 58.

Hora, S. L 1940 Dams and the problems of migratory fishes. *Curr. Sci.*, 9(9): 406–407.

Hora, S. L.1941 Life history and wanderings of hilsa in Bengal waters. *J.Asiat. Soc. (Sci.)*, 6(2): 93–112.

Hora, S. L. and K. K. Nair 1940a Further observations on the bionomics and fishery.

Hora, S. L. and K. K. Nair 1940b The Jatka fish of Eastern Bengal and its significance in the fishery of the so–called Indian shad, Hilsa ilisha (Hamilton). *Ibid.*, 42(4): 553–565.

Jones, S. and K. H. Sujansingani 1951 The Hilsa fishery of the Chilka lake. *J. Bombay Nat. Hist. Soc.*, 50(2): 264–280.

Malhotra, J. C. and K. I. Shah 1979 Hilsa—its breeding and possibilities of culture. Souvenir in commemoration of the ICAR Golden Jubilee Year, CIFRI. Barrackpore, Pt. 2: 175–178.

Mojumdar, C. H. 1939a – Culture of Hilsa. *Mod. Rev.*, 293–295.

Naidu, M. R. 1934 Report on a survey of the fisheries of Bengal. Calcutta.

Pandit, C. G. and S. L. Hora 1951 The probable role of Hilsa, *Hilsa iisha* (Hamilton), in maintaining cholera endemicity in India. Indian *J. Med. Sci.*, 15(7): 343–356.

Pantulu, V. R. 1966 Contributions to the study of the biology and fishery of some estuarine fishes. D. Phil. Thesis, Calcutta University.

Pillay, T. V. R. 1958a Biology of the hilsa, *Hilsa iisha* (Hamilton) of the river Hooghly. Ibid., 5: 201 –257.

Pillay, T. V. R. 1958b Morphological and physiological characters of the blood of hilsa, Hilsa iisha (Hamilton) of the river Hooghly. *Proc. Indian Acad. Sci*, B47(3): 155–162.

Pillay, S. R. and K. V. Rao 1963 Observations on the biology and fishery of hilsa, *Hilsa ilisha* (Hamilton), of river Godavari. Proc. Indo–Pac/f. Fish. Coun., 10(2): 37–61.

Pillay, S. R., K. V. Rao and P. K. Mathur 1963 Preliminary report on the tagging of the hilsa, Hilsa ills/ia (Hamilton). *Ibid.*, 10(2): 28–36.

Pillay, S. R. and H. Rosa Jr. 1963 Synopsis of biological data on hilsa, *Hilsa ilisha* (Hamilton) 1882. FAQFish B/al. Synops., (25): 1: 1 –6 :38.

Prashad, B., S. 1. Hora and K. K. Nair 1940 Observations on the seaward migration of the so–called Indian shad, H/Isa ilisha (Hamilton). *Rec. Indian Mus.*, 42(4): 529—552.

Rajyalakshmi, T. 1973 The population characteristics of the Godavari hilsa over the years 1963–1967. Indian J. Fish., 20(1): 78–94.

Regan, 1917. A revision of the clupeiod fishes of the genera *Pomolobus, Brevoorita and Dorosoma* and their llies. *Ann. Mag. Nat. Hist.*, 19: 304.

Russel, 1803. Description and figures of two hundred fishes collected at Visakhapatnam on the coast of Coromandal. W. Bulmer and Co., London, pp: 77–78.

Southwell, T. and B. Prashad 1918 On Hilsa investigations in Bengal, Bihar and Orissa. *Ibid.*, 11: 1–12.

Chapter 3

Bioremediation: A Microbial Technology for Improvising Wildlife

Anita Sharma and Pankaj Bhatt

Department of Microbiology,
College of Basic Sciences and Humanities, GBPUAT
Pantnagar – 263 145, Uttarakhand

ABSTRACT

Increased human interference in natural ecosystems has drastically affected microbial biodiversity, flora/fauna and animal lives. Impact of interference is severe and could be observed in the form of changed seasonal pattern. Increased global temperature is affecting our cropping systems, glaciers, urban and wild life. Increased level of green gases has affected physicochemical properties of soil, water and air quality. Microorganisms are integral part of biogeochemical cycles, internal body parts of the animals and plant tissues. Microbial diversity plays a key role in maintenance of geochemical cycles through different biological processes. Xenobiotics are man made compounds and reported to be significant contributors in global economy by way of protection of crops and human health. They are not easily attacked by natural microbial flora and are found to be major cause of environmental pollution. Concentration of such compounds is increasing day by day in air, soil and water bodies. In the process of biomagnifications, these compounds are badly affecting human health, fishes and animals. In this review we are focusing the destruction or disturbance of wildlife due to hazardous compounds. Emphasis on reclamation of soil/water systems with the help of potent microbes has also been undertaken.

Keywords: *Bioremediation, Microbial, Xenobiotic, Waste.*

Introduction

Wildlife includes living things that are neither human nor domesticated, especially mammals, birds and fishes hunted by man. Wildlife also includes all plants, fungi and microorganisms which grow or live wildly in an area without being introduced by humans. Biodiversity term is used for the variety of species of all living plants, animals and microbes living in their natural habitats. Domesticating wild plants and animal species for human benefit has occurred many times all over the world, and has a major impact on the environment and its component. Deserts, forests, rain forests, plains, grasslands, and other areas including developed urban cities have distinct forms of wildlife. The wildlife of these sites is also associated with microbial flora and fauna.

Environmental pollution significantly affects ecology of the nature. It is related to degree of pollution which is directly concerned with degree of change in the natural systems. Change in the natural systems affects wild life, existing in that ecosystem. The loss of wild life due to increased use of xenobiotic compounds in the environment is also a major problem of today's world. Bioremediation of such pollutants through microbial technology could be one of the alternative ways to restore the wild life from contaminated environment. The role of the microorganisms to remediate the polluted environment is great and could be related to their wide variety of metabolic activities, presence and high reproducing rate. Bioremediation is a waste management technique that involves the use of organisms to remove or neutralize pollutants from a contaminated site. These technologies help to remediate the polluted environment and help in maintenance of wild life by protecting them from various diseases. Fish is a major source of food in many regions of the world. Hazardous chemicals, coming from different sources accumulate in the ground water are taken up by the fish and finally utilized by the top consumer of the food chain. Extensive use of pesticides and plant protectors cause imbalance in the physicochemical properties of soil, water and air environments. Such chemicals create diverse environmental problem via biomagnifications. Currently, microbial degradation is one of the important techniques for amputation and degradation of pesticide from agricultural soils. Some recent studies have reported the use of genetically modified microorganisms in biodegradation of hazardous wastes/ chemicals under controlled conditions but introduction of these microbes in the fields may cause other environmental problems. Microbial consortia of indigenous and naturally occurring microbes isolated from particular contaminated environment show the ability to degrade pesticides at faster rate. Bioaugumentation processes like addition of necessary nutrients or organic matter are required to speed up the rate of degradation of a contaminant by the indigenous microbes. The use of indigenous microbial strains having plant growth promotory activities is ecologically superior technology of bioremediation of polluted sites over the chemical methods (Verma 2014).

Effect of Xenobiotic Compounds on Diversity of Wildlife

Different man-made xenobiotics are being used to kill phytopathogens for enhanced crop yield in every corner of agriculture systems. Of the three billion kg of pesticides produced annually, more than 75 per cent is used only in agricultural practices (Moore, 2007). The first generation of pesticides, introduced over 40 years ago, included organochlorine compounds and organophosphorus. Seventy-five per cent of the total pesticides produced globally is being consumed by developed countries. Developing countries are also showing increasing trend in pesticide consumption (Miller, 2004). Despite the benefits of pesticides in reducing the proliferation of pests, these chemicals have shown adverse health and ecological impacts. Due to improper storage conditions, man handling, spill and dissipation, pesticides are posing serious environmental, health and safety problems. These xenobiotics released in environment affect all the animals, plants and microbial communities in natural environment. In animals, these compounds mainly affect the nervous system, endocrine system by which the gene expression is changed. Due to increase in the concentration, various Xenobiotics majorly cause of wildlife destruction.

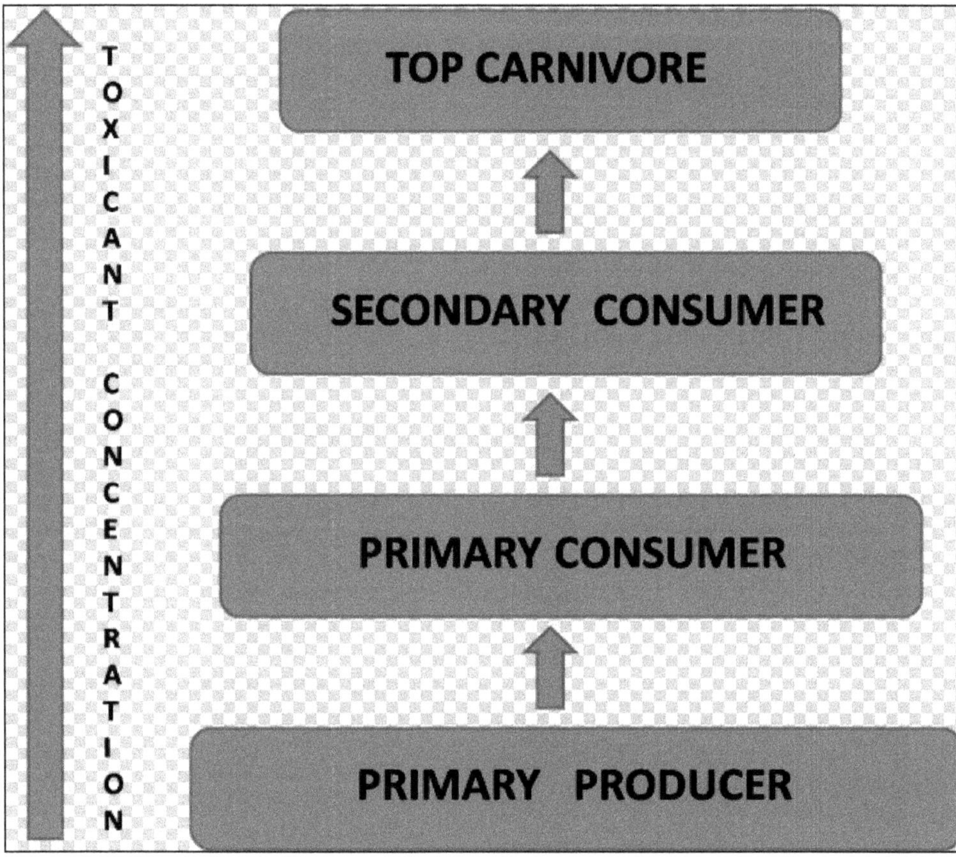

Impact of Pesticide on Vertebrate Wildlife.

There are many xenobiotics that affect the wildlife and pesticide is one of them. Although vertebrates are the intended target of only 2 per cent of pesticides available in the market, the unintentional impacts of pesticides on vertebrate populations have been marked and are well documented (Sánchez-Bayo 2011). Pesticides can exert their impact on vertebrates either directly (through their toxicity) or indirectly, for example, by reducing their food supply. Direct effects may be the result of several different exposure pathways: through ingestion of the formulated product *e.g.* birds eating seeds coated with insecticide (Avery *et al.*, 1997; Prosser and Hart 2005), through uptake via the skin following a spray event (Mineau 2011) or by eating contaminated prey. Depending on the extent of intoxication, pesticides can either kill vertebrates outright or exert sub-lethal effects on their growth and reproduction. Progress since the organochlorine era, has helped ensure that compounds that are currently being developed and registered are generally less persistent and do not readily bio-accumulate in food webs and are less toxic. Over the last 2 decades, a new class of insecticides, the neonicotinoids, has become the most important and fastest growing of the five major chemical classes of insecticides used in the global market (Jeschke and Nauen 2008; Jeschke *et al.*, 2011; Tomizawa and Casida 2011; Casida and Durkin 2013). When used as plant protection products, neonicotinoids act by becoming distributed systemically throughout the growing plant following seed or soil applications. Another recent insecticide, fipronil, a phenyl-pyrazole (fiprole) rather than a neonicotinoid, also acts in the same manner and has a similar toxicity and persistence profile (Grant *et al.*, 1998). Consequently, the neonicotinoids and fipronil are sometimes jointly termed 'systemic insecticides', although there are other products which could be termed 'systemic', for example, acephate (the organo-phosphorus insecticide) and the organoarsenical, monosodium methanearsonate. Neonicotinoids are, in particular, commonly applied as seed treatments. The use of seed treatments as a convenient and effective application method has widespread appeal in the farming industry. Consequently, systemic seed treatments are now used on the majority of agricultural crops worldwide (Garthwaite *et al.*, 2003; Jeschke *et al.*, 2011). Mode of action of the systemic insecticides Neonicotinoids work by interfering with neural transmission in the central nervous system. They bind to the nicotinic acetylcholine receptors (nAChR) in the postsynaptic neuron, acting as 'false neurotransmitters' (agonists). This interference with acetylcholine neurotransmitter signalling causes continuous activation of the receptor, leading to symptoms of neurotoxicity. Neonicotinoids have greater affinity for, and thus bind more strongly to, insect than mammalian or other vertebrate receptors, so their toxicity to mammals is lower than to insects and the reversibility of intoxication higher (Tomizawa and Casida 2005; Jeschke *et al.*, 2011). Fipronil works similarly, but instead binds to the gamma-aminobutyric acid (GABA) receptors, resulting in similar continuous central nervous system activity (Tingle *et al.*, 2003). Fipronil has lower affinity with vertebrate than to invertebrate receptors (**Gibbons *et al.*, 2014).**

Systemic Insecticides and their Impact on Ecosystem Functioning and Services

Large-scale use of the persistent and potent neonicotinoid and fipronil insecticides has raised concerns about risks to ecosystem functions provided by a wide range of species and environments affected by these insecticides (Chignon *et al.*, 2014). The concept of ecosystem services is widely used in decision making in the context of valuing the service potentials, benefits, and use values that well-functioning ecosystems provide to humans and the biosphere and, as an endpoint (value to be protected), in ecological risk assessment of chemicals. Neonicotinoid insecticides are frequently detected in soil and water and are also found in air, as dust particles during sowing of crops and aerosols during spraying. These environmental media provide essential resources to support biodiversity, but are known to be threatened by long-term or repeated contamination by neonicotinoids and fipronil.

Empirical studies examining the specific impacts of neonicotinoids and fipronil to ecosystem services have focused largely on the negative impacts to beneficial insect species (honeybees) and the impact on pollination service of food crops. In particular, microbes, invertebrates, and fish play critical roles as decomposers, pollinators, consumers, and predators, which collectively maintain healthy communities and ecosystem integrity. Negative impacts of systemic insecticides on decomposition, nutrient cycling, soil respiration, and invertebrate populations valued by humans have also been assessed. Invertebrates, particularly earthworms that are important for soil processes, wild and domestic insect pollinators which are important for plant and crop production, and several freshwater taxa which are involved in aquatic nutrient cycling, are found to be highly susceptible to lethal and sublethal effects of neonicotinoids and/or fipronil at environmentally relevant concentrations. By contrast, most microbes and fish do not appear to be as sensitive under normal exposure scenarios, though the effects on fish may be important in certain realms such as combined fish-rice farming systems and through food chain effects.

Conservation of Natural Enemy Populations in Greenhouse Crops: Current Methods and Future Prospects

Biological control of arthropod pests has a long tradition in greenhouse crops. Both the area on which it is used and the number of available biological control agents are still expanding (Pilkington *et al.*, 2010; van Lenteren 2012). Biological control programmes in greenhouses are often based on periodical releases of natural enemies, also referred to as augmentative biological control (van Lenteren 2012). Although biological control has proven to be successful in many greenhouse crops, but efficacy can be insufficient in ornamentals plants (Heinz *et al.*, 2004). Poor establishment and persistence of natural enemies in certain crops can be one of the main problems in pest control, which is partly due to the types of natural enemies used. The selection of natural enemies for augmentative biological control was traditionally focused on specialist natural enemies that were released to obtain rapid control of the pests (van Lenteren and Woets, 1988). Well-known examples are the spider mite predator *Phytoseiulus persimilis*, Athias- Henriot, the whitefly parasitoid Encarsia Formosa Gahan and the aphid parasitoids *Aphidius* spp., which are still successfully, used, in many crops and countries (van Lenteren, 2012). Although, these specialists are well adapted to their host and can be very effective, they often disappear when prey density is reduced. As they are used mainly to obtain rapid control of specific pests, their efficacy requires high quantity and quality of released natural enemies and intensive monitoring to assure accurate timing of the intervention. To overcome the problem of establishment and monitoring, some specialist natural enemies (*e.g.* aphid and whitefly parasitoids) are released routinely (*e.g.* weekly) as an "insurance policy". However, this method is not always economically viable. Methods that increase the persistence of natural enemies in crops could greatly enhance the efficacy, robustness and cost-effectiveness of biological pest control. The establishment and persistence of generalist predators compared to specialist natural enemies may provide more sustainable biological control, as their broader diet range enables them to persist or even reproduce on alternative prey or plant-provided food sources in the absence of pest organisms

(Symondson *et al.*, 2002). This offers the opportunity to inoculate crops that provide such food sources with generalist predators before pest invasions (preventive biological control). However, many crops do not provide the additional resources required by natural enemies. Impediments to successful establishment include: insufficient plantprovided food, or plant-provided food of insufficient quality; lack of suitable oviposition sites; lack of shelter and absence of prey. Biological control might be enhanced in such crops by supplementing the missing resources and thus providing conditions that facilitate more successful establishment of natural enemies. Conservation of naturally occurring natural enemies (conservation biological control) is well developed in outdoor crops where various techniques of habitat modifications are used such as flowering strips, cover crops that provide windborne pollen ormulching (Landis *et al.*, 2000; Maoz *et al.*, 2011; Wa¨ckers and van Rijn 2012). Biological control in greenhouse crops might be enhanced by using similar methods, but the cost intensive production of many greenhouse crops means that conservation methods that compromise valuable cropping areas are usually not feasible. Preventive biological control through conservation techniques may help overcome many problems of greenhouse biological control, like the issues of adequate timing, pest detection, the high quantity of natural enemies required, and the labour and knowledge requirements (Mullens *et al.*, 2008).

Establishment and spread of *Ornithonyssus sylviarum* were documented through time on sentinel hens (50 per house of 28,000–30,000 hens) in the first egg production cycle of three large commercial flocks (12 houses) of white leghorn hens. Mites were controlled using acaricide, and the impacts of treatment on mite populations and economic performance were documented. Mite prevalence and intensity increased rapidly and in tandem for 4–8 weeks after infestation. Intensity declined due to immune system involvement, but prevalence remained high, and this would affect mite sampling plan use and development. Early treatment was more effective at controlling mites; 85 per cent of light infestations were eliminated by a pesticide spray (Ravap), versus 24 per cent of heavy infestations. Hens infested later developed lower peak mite intensities, and those mite populations declined more quickly than on hens infested earlier in life. Raw spatial association by distance indices (SADIE), incorporating both the intensity and distribution of mites within a house, were high from week-to-week within a hen house. Once adjusted spatially to reflect variable hen cohorts becoming infested asynchronously, this analysis showed the association index tended to rebound at intervals of 5–6 weeks after the henimmune systemfirst suppressed them. Large, consistent mite differences in oneflock (highvs. lowinfestation levels) showed the economic damage of mite parasitism (assessed by flock indexing) was very high in the initial stages of mite expansion. Unmitigated infestations overall reduced egg production (2.1–4.0 per cent), individual egg weights (0.5–2.2 per cent), and feed conversion efficiency (5.7 per cent), causing a profit reduction of $0.07–0.10 per hen for a 10-week period. Asynchronous infestation patterns among pesticide-treated hens may have contributed to a lack of apparent flock-level economic effects later in the production cycle. Individual egg weights differed with mite loads periodically, but could be either higher or lower, depending on circumstances and interactions with hen weight. Individual hen weight gains were depressed by high/moderate mite loads, but the heavier hens in a flock harbored more mites. This led to compensatory weight gains after mites declined. Trade offs between resource allocation to body growth or production versus immune system function appeared to be operating during the early and most damaging mite infestation period, when high egg production was beginning and the hens were gaining weight. The results were related to other studies of mite impact on domestic hens and to wild bird–ectoparasite studies. Much of the mite economic damage probably is due to engaging and maintaining the immune response.

Bioremediation Methods

In situ and *ex situ* methods Bioremediation technologies can be broadly classified as *ex situ* and *in situ*. *Ex situ* technologies are those treatments which involve the physical removal of the contaminated material for treatment process. In contrast, in situ techniques involve treatment of the contaminated material in place. Some of the examples of *in situ* and *ex situ* bioremediation are given below:

1. Land farming: Solid-phase treatment system for contaminated soils: may be done in situ or ex situ.
2. Composting: Aerobic, thermophilic treatment process in which contaminated material is mixed with a bulking agent; can be done using static piles or aerated piles.
3. Bioreactors: Biodegradation in a container or reactor; may be used to treat liquids or slurries.
4. Bioventing: Method of treating contaminated soils by drawing oxygen through the soil to stimulate microbial activity.
5. Biofilters: Use of microbial stripping columns to treat air emissions.
6. Bioaugmentation: Addition of bacterial cultures to a contaminated medium; frequently used in both *in situ* and *ex situ* systems.
7. Biostimulation: Stimulation of indigenous microbial populations in soils or ground water by providing necessary nutrients.
8. Intrinsic bioremediation: Unassisted bioremediation of contaminant; only regular monitoring is done.
9. Pump and treat: Pumping ground water to the surface, treating, and reinjecting.

Bioremediation Processes Carried Out by Microorganisms

Microorganisms and Methane Oxidation

Traditionally, studies on pollutant biodegradation are initiated by the isolation of one or more microorganisms capable of degrading target pollutants; however, conventional isolation methods mostly result in the isolation of only a fraction of the diverse pollutant-degrading microorganisms in the environment. In addition, most isolated organisms show pollutant-degradation kinetics that differ from those observed in the environment. For example, laboratory-cultivated methanotrophs exhibit half-saturation constants for methane oxidation which are one to three orders of magnitude higher than those observed in soil. Using molecular phylogenetic analyses of isotope-labeled DNA, **Radajewski *et al.* (2000)** successfully identified two novel methanotrophs that actively degrade methane under environmental conditions. Methanotrophs are considered to be important for reducing the emission of methane, a greenhouse gas, from soil and sediment. In addition, methanotrophs co-metabolize trichloroethylene (TCE); therefore, TCE bioremediation often employs methane injection as a means to stimulate the TCE-degrading activity of indigenous methanotrophs (*i.e.* methane biostimulation). Methanotrophs which occur at a methane biostimulation site can be analyzed using denaturing gradient gel electrophoresis (DGGE) of polymerase chain reaction (PCR)-amplified 16S rDNA and soluble methane monooxygenase gene fragments.

Marine Petroleum Hydrocarbon Degradation

Spilled-oil bioremediation experiments, conducted at a sandy beach found that phylotypes affiliated with the α subclass of *Proteobacteria* appeared in the DGGE fingerprints obtained for oiled plots but not in those for unoiled plots (**MacNaughton *et al.*, 1999)**, suggested their importance in spilled-oil bioremediation. Another oil-spill experiment conducted at a beach in the Norwegian Arctic showed that 16S rDNA types affiliated with the γ-*Proteobacteria*, especially those belonging to the *Pseudomonas* and *Cycloclasticus* groups, were abundant in fertilized oil sands **(Grossman *et al.*, 2000)**. Microbial populations which occurred in seawater after supplementation with petroleum and inorganic fertilizers have been analyzed using rRNA approaches; it was reported that bacterial populations belonging to the α-*Proteobacteria* and the genus *Alcanivorax* **(Chang *et al.*, 2000; Harayama *et al.*, 1999)** showed accelerated growth. These studies have indicated that some groups of bacteria commonly occur in oil-contaminated marine environments, although other populations change under different environmental conditions.

Anaerobic Degradation of Petroleum Hydrocarbons

As petroleum hydrocarbons are persistent under anaerobic conditions, their contamination of groundwater is a serious environmental problem. The microbial diversity in a hydrocarbon and chlorinated-solvent contaminated aquifer undergoing intrinsic bioremediation can be assessed by cloning and sequencing bacterial and archaeal 16S rDNA fragments. This study detected phylotypes that were closely related to *Syntrophus* spp. (anaerobic oxidizers of organic acids with the production of acetate and hydrogen) and *Methanosaeta* spp. (aceticlastic methanogens), suggesting their syntrophic association. Phylotypes affiliated with candidate divisions (that do not contain any isolated organisms) were also obtained in abundance from the contaminated aquifer **(Dojka *et al.*, 1998)**, although their physiology is completely unknown. A similar syntrophic association of bacteria and archaea has also been reported in a methanogenic enrichment that slowly degrades hexadecane **(Zengler *et al.*, 1999)**. Likewise, a toluene-degrading methanogenic consortium was characterized by rRNA approaches **(Ficker *et al.*, 1999)**.

Biodegradation of Polycyclic Aromatic Hydrocarbons

Polycyclic aromatic hydrocarbons (PAHs) are compounds of intense public concern owing to their persistence in the environment and potentially deleterious effects on human health. A soil-derived microbial consortium capable of rapidly mineralizing benzo[a]pyrene was analyzed by DGGE profiling of PCR-amplified 16S rDNA fragments **(Kanaly *et al.*, 2000)**. The analysis represented organisms closely related to known high molecular weight PAH-degrading bacteria (*e.g. Burkholderias, Sphingomonas* and *Mycobacterium*). In soil environments, the reduced bioavailability of PAHs due to sorption to natural organic matter is an important factor controlling their biodegradation. Different phenanthrene-degrading bacteria occurred in soil enrichments when different sorptive matrices were present **(Friedrich *et al.*, 2000)**. It has also been shown that the application of surfactants to soil enrichments that degrade phenanthrene and hexadecane altered the microbial populations responsible for the degradation **(Colores *et al.*, 2000)**.

Metal Bioremediation

Because of its toxicity, metal contamination of the environment is also a serious problem. Studies have applied molecular tools to the analysis of bacterial and archaeal populations that are capable of surviving in metal-contaminated environments. Bacterial communities in soil amended for many years with sewage sludge that contained heavy metals were assessed using rRNA approaches, including FISH and cloning and sequencing **(Sandaa *et al.*, 1999)**. The study found that two sequence groups affiliated with the α-*Proteobacteria* and *Actinobacteria* were frequently obtained from clone libraries from the metal-contaminated soil, although most *Actinobacteria* sequences showed low similarity (<85 per cent) to the sequences of any hitherto cultured actinomycete. The detoxification machineries that some of these organisms may have are considered useful for metal bioremediation, and comparisons with the machineries of previously isolated metal-resistant bacteria may yield interesting results. Heavy-metal-tolerant *Ralstonia eutropha* can be genetically engineered to express mouse metallothionein on the cell surface **(Valls *et al.*, 2000)**. It was demonstrated that the inoculation of Cd^{2+}-polluted soil with the genetically engineered *Ralstonia* significantly decreased the toxic effects of the heavy metal on the growth of tobacco plants.

Waste Treatment

Microbial consortia involved in wastewater treatment have been a major subject of microbial ecology, and many papers have been published in which molecular tools were used for community analyses. Bacterial community structures and physiological states within an industrial phenol bioremediation system can be analyzed.

Comparisons made between the amounts of group-specific rRNAs and the process chemistry can help to identify some phylogenetic groups of bacteria important for the process performance. The phylogenetic diversity of bacterial communities supported by a seven-stage, full scale bioreactor used to treat pharmaceutical wastewater was studied using PCR-based techniques (*i.e.* DGGE fingerprinting and cloning of 16S rDNA fragments)(**LaPara *et al.*, 2000)**. A combination of 16S rDNA cloning, hybridization with oligonucleotide probes for ammonia-oxidizing bacteria (AOB) and sequencing of the hybridization-positive clones suggested that novel *Nitrosospira*-like populations were the major AOB in a rhizosphere zone used to treat wastewater (rhizoremediation) **(Abd El Haleem *et al.*, 2000)**. To identify microbial populations responsible for phosphorus removal in activated-sludge, the structure of the bacterial population was analyzed by FISH during the operation of a laboratory-scale reactor with various phosphorus removal rates **(Bond *et al.*, 1999)**. FISH has also been used to analyze microbial populations in mesophilic and thermophilic sludge granules, foaming activated-sludge and bulking activated-sludge. Temperature-gradient gel electrophoresis (TGGE) of PCR-amplified 16S rDNA fragments can be used to identify the major phylotypes in phenol-digesting activated-sludge.

Bioaugmentation

The introduction of exogenous microorganisms into environments (bioaugmentation) has been used in an attempt to accelerate bioremediation. It is desirable that the fate of an introduced organism be monitored in order to prove its contribution to pollutant degradation and to assess its influence on the ecosystem. Molecular tools have been used for this purpose. DGGE/TGGE fingerprinting of 16S rDNA fragments has been used to examine the effects of bioaugmentation on indigenous bacterial community structures in a range of situations: a laboratory-scale semi-continuous activated-sludge system loaded with 3-chloroaniline (Boon *et al.*, 2000). In some cases, where genetically modified organisms were utilized, bioaugmentation improved pollutant-biodegradation rates in the environment due to the establishment of transconjugants capable of degrading the pollutants rather than the direct contribution of the inoculated organisms.

Advantages and Disadvantages of Bioremediation

For bioremediation to be successful, the bioremediation methods depend on having the right microbes in the right place with the right environmental factors for degradation to occur. The right microbes are bacteria or fungi, which have the physiological and metabolic capabilities to degrade the pollutants. Bioremediation offers several advantages over conventional techniques such as land filling or incineration. Bioremediation can be done on site, is often less expensive and site disruption is minimal, it eliminates waste permanently, eliminates long-term liability, and has greater public acceptance, with regulatory encouragement, and it can be coupled with other physical or chemical treatment methods. Bioremediation has also its limitations. Some chemicals are not amenable to biodegradation, for instance, heavy metals and some chlorinated compounds. In some cases, microbial metabolism of contaminants may produce toxic metabolites. Bioremediation is a scientifically intensive procedure which must be tailored to the site-specific conditions, which means one has to do treatability studies on a small scale before the actual clean-up of the sites.

Some of the questions one has to answer before using bioremediation techniques are: is the contaminant biodegradable? is biodegradation occurring in the site naturally? are environmental conditions appropriate for biodegradation? if the waste does not completely biodegrade, where will it go? These questions can be answered by doing site characterization and also by treatability studies.

Conclusion

Xenobiotic compounds are increased in our surrounding environment every day. These compounds mainly affect the living system biology. Till date many xenobiotics reported as a carcinogenic compound. Biomagnification of these compounds affect the wildlife at a particular trophic level. So complete removal/ degradation of such compounds is possible with the microorganisms. Microbial system produces extracellular and intracellular enzymes for the degradation of such types of compounds from environment. By using the potential microorganisms we can remove the hazardous compounds from the trophic level, by which the rapid reduction of such compound is possible in animal system. So such type of approach to reduce the biomagnifications of pollutant is important for protection of wild animal and plant species from hazardous chemicals.

References

Abd El Haleem, D., von Wintzingerode, F., Moter, A., Moawad, H. and Gobel, U.B. 2000. Phylogenetic analysis of rhizosphere-associated β-subclass proteobacterial ammonia oxidizers in a municipal wastewater treatment plant based on rhizoremediation technology. *Lett Appl Microbiol.* 31:34-38.

Bond, P.L., Erhart, R., Wagner, M., Keller, J. and Blackall, L.L. 1999. Identification of some of the major groups of bacteria in efficient and nonefficient biological phosphorus removal activated sludge systems. *Appl Environ Microbiol.* 65:4077-4084.

Boon, N., Goris, J., De Vos, P., Verstraete, W. and Top, E.M. 2000. Bioaugmentation of activated sludge by an indigenous 3-chloroaniline-degrading *Comamonas testosteroni* strain, I2gfp. *Appl Environ Microbiol.* 66:2906-2913.

Chang, Y.J., Stephen, J.R., Richter, A.P., Venosa, A.D., Bruggemann, J., MacNaughton, S.J., Kowalchuk, G.A., Haines, J.R., Kline, E. and White, D.C. 2000. Phylogenetic analysis of aerobic freshwater and marine enrichment cultures efficient in hydrocarbon degradation: effect of profiling method. *J Microbiol Methods.* 40:19-31.

Colores, G.M., Macur, R.E., Ward, D.M., Inskeep, W.P. 2000. Molecular analysis of surfactant-driven microbial population shifts in hydrocarboncontaminated soil. *Appl Environ Microbiol.* 66:2959-2964.

Dojka, M.A., Hugenholtz, P., Haack, S.K and Pace, N.R. 1998. Microbial diversity in a hydrocarbon- and chlorinated-solvent-contaminated aquifer undergoing intrinsic bioremediation. *Appl Environ Microbiol.* 64:3869-3877.

Ficker, M., Krastel, K., Orlicky, S. and Edwards, E. 1999. Molecular characterization of a toluene-degrading methanogenic consortium. *Appl Environ Microbiol.* 65:5576-5585.

Friedrich, M., Grosser, R.J., Kern, E.A., Inskeep, W.P. and Ward, D.M. 2000. Effect of model sorptive phases on phenanthrene biodegradation: molecular analysis of enrichments and isolates suggests selection based on bioavailability. *Appl Environ Microbiol.* 66:2703-2710.

Grossman, M.J., Prince, R.C., Garrett, R.M., Garrett, K.K., Bare, R.E., O'Neil, K.R., Sowlay, M.R., Hinton, S.M., Lee, K. and Sergy, G.A. 1999. Microbial diversity in oiled and un-oiled shoreline sediments in the Harayama S, Kishira H, Kasai Y, Shutsubo K: Petroleum biodegradation in marine environments. *J Mol Microbiol Biotechnol.* 1:63-70.

Kanaly, R.A., Bartha, R., Watanabe, K. and Harayama, S. 2000. Rapid mineralization of benzo[a]pyrene by a microbial consortium growing on diesel fuel. *Appl Environ Microbiol.* 66:4205-4211.

LaPara, T.M., Nakatsu, C.H., Pantea, L. and Alleman, J.E. 2000. Phylogenetic analysis of bacterial communities in mesophilic and thermophilic bioreactors treating pharmaceutical wastewater. *Appl Environ Microbiol.* 66:3951-3959.

MacNaughton, S.J., Stephen, J.R., Venosa, A.D., Davis, G.A., Chang, Y.J. and White, D.C. 1999. Microbial population changes during bioremediation of an experimental oil spill. *Appl Environ Microbiol.* 65:3566-3574.

Radajewski, S., Ineson, P., Parekh, N.R., Murrell, J.C. 2000. Stable-isotope probing as a tool in microbial ecology. *Nature.* 403:646-649.

Sandaa, R., Torsvik, V., Enger, O., Daae, F.L., Castberg, T. and Hahn, D. 1999. Analysis of bacterial communities in heavy metal-contaminated soils at different levels of resolution. *FEMS Microbiol Ecol.* 30:237-251.

Valls, M., Atrian, S., de Lorenzo, V., Fernandez, L.A. 2000. Engineering a mouse metallothionein on the cell surface of *Ralstonia eutropha* CH34 for immobilization of heavy metals in soil. *Nat Biotechnol.* 18:661-665.

Zengler, K., Richnow, H.H., Rossello-Mora, R., Michaelis, W. and Widdel, F. 1999. Methane formation from long-chain alkanes by anaerobic microorganisms. *Nature.* 401:266-269.

Chapter 4

An Overview of Indian Wildlife

Nidhi Mishra

College of Fisheries,
GBPUA&T, Pantnagar, Uttarakhand

ABSTRACT

Wildlife is considered as natural wealth of a country.Hence, it needs to be conserved and optimally utilized for social development. Unmindful human activities threaten the existence of wildlife.This chapter discusses the reasons for endangered animals.List of endangered species, strategies to conserve them and national and international agencies involved in conservation of Wildlife.

***Keywords**: Conservation, Endangered, Exploitation, Wild.*

Introduction

The Indian subcontinent boasts of serving as the natural habitat of a large and varied wildlife species possibly due to the broad range of weather, soil, climate, and other similar factors. We can find some of the most magnificent as well as the rarest wildlife species of the world in the country. The beauty and variety we see in the jungles of India is difficult to be expressed in words. Wildlife includes wild animals and plants collectively. As a human being, we need to live in amity with other living things on the Earth. On the other side, however, there are certain species that are reported threatened or vulnerable and are required protection. Humans should be blamed causing this problem occurs. Cutting down the trees in jungles destroys habitats of wildlife. As a result, wildlife loses their food and hence becomes endangered species. This problem leads to imbalance of ecosystem. Perhaps the need of the time is to put more and more conservation efforts to protect wildlife in India.

India is developing in many ways, every day. But this development is giving antagonistic results for Indian wildlife. Developing cities, villages have resulted in a rapid decline in the population of most of the wild animals and birds. We are killing the innocent angelic beings just for our sake. There are many reasons of ending wildlife, not only developing cities, villages and construction, hunting also does the same. Hunting is done for self-consumption or for merchandize fleshes, horns, antlers, pelt, bones, feathers, etc worldwide for cash income.

Species have been ending, since this development began. We have reached a point when a large percent of the species have vanished. Wildlife includes plants, trees, birds and many others organisms. Wildlife is the major reason of human's endurance. Wildlife of India has variety of organisms of different origins. The region's divergent wildlife is preserved in plentiful national parks and wildlife sanctuaries across the country. Since India is a centre of large number of rare and blustered animal species, wildlife management in the country is indispensible to keep it safe and sheltered.

Conservation of Indian wildlife was not given the requisite importance for a long time. The past few decades have seen the greed and negligence of human beings working to the detriment of this rich wildlife. However, the government as well as the people slowly and gradually understood their responsibility in this context. In 1972, India started the consummate Wildlife protection Act and Project tiger to bulwark crucial habitat. Hunting became illegal in India in 1972, after the commencement of Wildlife Protection Act, with the essence to save wildlife in India. Many foundations in India are working on it, but somehow evils of society are doing their work, that is eradicating of the ecofriendly environment. Actually wildlife is daily moving a step ahead to be vanished.

View on Wildlife

India's rich and abundant animal life is attributed to its diverse climate and geography. Of the total 63,837 species globally assessed, the IUCN classified 3,947 as Critically Endangered, 81 as Extinct, 63 as Extinct in the Wild. The richness of wildlife can be estimated in India with the fact that it is home to about 7.6 per cent of all mammal population, 12.6 per cent of avian, 6.2 per cent of reptiles and 6.0 per cent of entire flowering species found on the Earth. The Red list of threatened species, prepared by the International Union for Conservation of Nature (IUCN), has listed 132 species of plants and animals as Critically Endangered, the most threatened category, from India. The Critically Endangered list included 18 species of amphibians, 14 fishes and 10 mammals. There are also 15 bird species in the category. The agency listed 310 species as endangered ones, including 69 fishes, 38 mammals and 32 amphibians. Two plant species were reported to be extinct in the wild. The richness of biological diversity in the India and world is shown in Figure 4.1.

Why are Animals Endangered?

An animal is considered to be endangered when it seems to disappear within next twenty years or so without any conservative measurements. Referring to an estimate, about 99.9 per cent of all plants and animals species have become extinct. Wildlife extinction results from many forces operating in the society, such

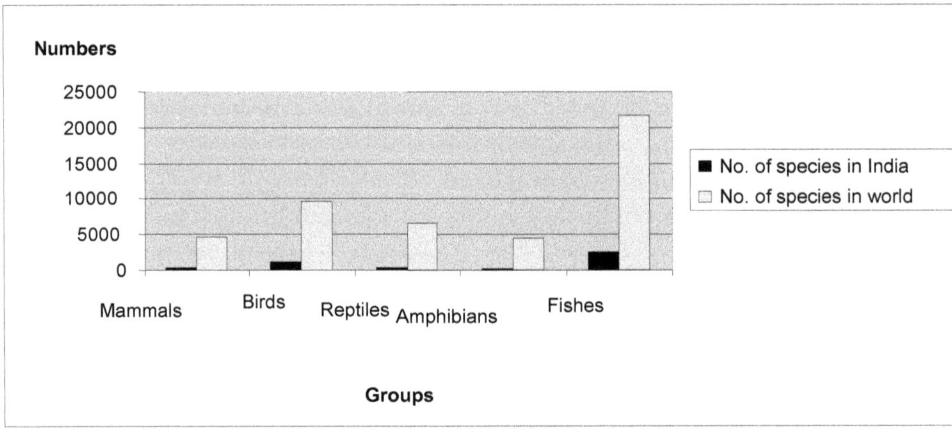

Figure 4.1: Total No. of Species in World and India.

as economics, politics and psychology, the specific activities that cause extinction of species and the relative importance of each.

However, today the biggest concern is not the process of extinction, but the rate at which it is increasing, with the role of humans at the forefront. Here are some reasons why animals become endangered

1. Habitat Loss

The primary cause of endangerment is the loss of habitat or native lands of many animal species. All animals require water, food, and shelter to live in, just as we humans do. Humans are, however, extremely adaptable for they can generate food, preserve water, and seek shelter on their own from raw materials. Animals cannot do this. Certain animals are extremely specialized in their dietary and habitat requirements as compared to others. These animals are likely to be endangered soon after the loss of their habitats. However, some animals are highly adaptable as they rely on more than one habitat to survive.

2. Exploitation

Many animals experience direct exploitation as a result of which there is a considerable decline in their numbers. There are some places where humans were mainly responsible for extreme hunting and poaching to meet their nutritional and commercial requirements altogether. When hunters ignore governmental rules that regulate the number of animals that should be hunted, they reduce populations to the point that species become endangered. Many animals were hunted to extinction.

3. Disturbance

As humans advance in machines, more and more animals begin to leave their natural habitats even if habitats are not damaged. However, still it is not clear as to whether humans played any role in a substantial wave of extinction in the prehistoric times.

4. Animals with Specialized Diet

Animals which rely on a particular diet such as fruit-eaters and carnivores have to travel great distances in search of prey as compared to those of herbivores or omnivores. Therefore, it's not surprising that many carnivores are dominating the endangered animals list.

5. Low Reproductive Output

Some animals have really developed reproductive schemes that in part reflect their possibility of dying. Large animals that have a longer lifespan and have few natural predators normally produce only one offspring each year. This ensures an absolute balance between the number of dying animals and the number which female must produce. Thus, when humans attempt to change the natural balance between deaths and births by increasing mortality, numbers decline. Large animals which are predominantly held in captivity are unable to reproduce quickly as they have low birth rates. Therefore, it will take a considerable amount of time to recover the numbers even with full protection.

6. Habitat Specialist

Animals that tend to occupy a limited habitat space become vulnerable when their habitats are widely separated in space. In these conditions, they will begin to inhabit in small populations which are susceptible to the natural disasters.

7. Small Population Size

Animals with a smaller population are likely to be extinct as compared to those with larger populations. With the drop in numbers, animals are limited to few isolated populations which further increase the probability of becoming endangered.

8. Limited Geographic Distribution

Certain animals have restricted geographic distribution as a result of which they are becoming endangered whenever their natural habitats are disturbed by humans.

9. Economic and Commercial Worth

It goes without saying that animals which have real commercial worth for their furs, oil, meat, for using it in products, are rather more susceptible to hunting and poaching.

10. Pollution

All living things including animals are affected by environmental pollutants. Human activity is releasing harmful chemicals into the air, water and soil all over the planet. Many animals cannot adapt these chemicals changes in the environment and they die. Pollution alters the habitat of plants and animals and plays a significant role in extinction. Water pollution is especially harmful to the organisms living in estuaries and coastal zones. Toxic wastes entering the food chain can have devastating effect on their pollution. Insecticides and pesticides can also affect the plants and animals.

11. Introduction of Exotics

Exotics are species that are typically non-native to a specific area. Humans have always involved in introducing exotic species to a particular area which have adverse effects on those native species. Exotic species often have a predatory or competitive advantage over native species, which have been a part of a particular biological environment for centuries. Even though native species are well adapted to their surroundings, they may not be able to deal with species that closely compete with them for food or hunt in ways that native species has not developed defenses against. As a result, native species either cannot find enough food to survive or are killed in such numbers as to endanger survival as a species

Some Critically Endangered Species of India

According to the Red Data Book of International Union for Conservation of Nature (IUCN), there are 47 critically endangered species in India (As on 5 September 2011). The Red list of 2012 is out on 18 February 2012 and has listed 132 species of plants and animals as Critically Endangered from India. The International Union for Conservation of Nature (IUCN) has marked Critically Endangered (CR) to be the highest risk category given to any wild species. This means that the species population is so less that it can become extinct any time. There are five criteria that govern whether a species can be called CR or not. These are:

1. Populations have declined or will decrease, by greater than 80 per cent over the last 10 years or three generations.
2. Have a restricted geographical range.
3. Small population size of less than 250 individuals and continuing decline at 25 per cent in 3 years or one generation.
4. Very small or restricted population of fewer than 50 mature individuals.
5. High probability of extinction in the wild.

Here is an outlook on some of the most critically endangered species of India and this does not limited to land animals only, there are species also that are undergoing rapid population decline. The population of these animals has been considerably reduced over the past 50 to 100 years. Let's take a look at these critically endangered animals

(A) BIRDS

1. The Jerdon's Courser (*Rhinoptilus bitorquatus*)
2. The Forest Owlet (*Heteroglaux blewitti*)
3. The White-bellied Heron (*Ardea insignis*)
4. Red-headed Vulture (*Sarcogyps calvus*)
5. The Great Indian Bustard (*Ardeotis nigriceps*)
6. White-backed Vulture (*Gyps bengalensis*)
7. Slender-billed Vulture (*Gyps tenuirostris*)
8. Long-billed Vulture (*Gyps indicus*)

9. The Himalayan Quail (*Ophrysia superciliosa*)
10. The Bengal Florican (*Houbaropsis bengalensis*)
11. The beautiful Pink- headed Duck (*Rhodonessa caryophyllacea*)
12. The Sociable Lapwing (*Vanellus gregarious*)
13. The Spoon Billed Sandpiper (*Eurynorhynchus pygmeus*)
14. The Siberian Crane (*Grus leucogeranus*)
15. Baer's Pochard (*Aythya baeri*)

(B) MAMMALS

1. The Pygmy Hog (*Porcula salvania*)
2. Andaman White-toothed Shrew (*Crocidura andamanensis*)
3. Jenkin's Andaman Spiny Shrew (*Crocidura jenkinsi*)
4. Nicobar White-tailed Shrew (*Crocidura nicobarica*)
5. Kondana Rat (*Millardia kondana*)
6. The Large Rock Rat or Elvira Rat (*Cremnomys elvira*)
7. The Namdapha Flying Squirrel (*Biswamoyopterus biswasi*)
8. The Blackbuck (*Antilopecervicapra*)
9. The Malabar Civet (*Viverra civettina*)
10. The Sumatran Rhinoceros (*Dicerorhinus sumatrensis*)
11. The Javan Rhinoceros (*Rhinoceros sondaicus*)
12. Dhole (*Cuon alpines*)
13. Himalayan Wolf (*Canis himalayensis*)
14. Kashmir Stag (*Cervus canadensis hanglu*)
15. Sumatran Tiger (*Panthera tigris sumatrae*)
16. Asian Elephant (*Elephas maximus indicus*)
17. Bengal Tiger (*Panthera tigris tigris*)
18. Indian Elephant (*Elephas maximus indicus*)
19. Snow Leopard (*Panthera unical*)
20. Red Panda (*Ailurus fulgens*)
21. Asiatic Lion (*Panthera leo persica*)

(C) REPTILES

1. The Gharial (*Gavialis gangeticus*)
2. The Hawksbill Turtle (*Eretmochelys imbricata*)
3. The Leatherback Turtle (*Dermochelys coriacea*)
4. The Loggerhead Sea Turtle (*Caretta caretta*)
5. Four-toed River Terrapin or River Terrapin (*Batagur baska*)
6. Red-crowned Roofed Turtle or the Bengal Roof Turtle (*Batagur kachuga*)
7. Sispara Day Gecko (*Cnemaspis sisparensis*)

(D) AMPHIBIANS

1. The Anamalai Flying Frog (*Rhacophorus pseudomalabaricus*)
2. The Gundia Indian Frog (*Indirana gundia*)
3. The Kerala Indian Frog (*Indirana phrynoderma*)
4. The Charles Darwin's Frog (*Ingerana charlesdarwini*)
5. The Kottigehar Bubble-nest Frog (*Micrixalus kottigeharensis*)
6. The Amboli Bush Frog (*Pseudophilautus amboli*)
7. The Chalazodes Bubble-Nest Frog (*Raorchestes chalazodes*)
8. The Small Bush Frog (*Raorchestes chotta*)
9. The Green-eyed Bush Frog (*Raorchestes chlorosomma*)
10. The Griet Bush Frog (*Raorchestes griet*)
11. The Kaikatt's Bush Frog (*Raorchestes kaikatti*)
12. The Mark's Bush Frog (*Raorchestes marki*)
13. The Munnar Bush Frog (*Raorchestes munnarensis*)
14. The Large Ponmudi Bush Frog (*Raorchestes ponmudi*)
15. The Resplendent Shrub Frog (*Raorchestes resplendens*)
16. The Sacred Grove Bush frog (*Raorchestes sanctisilvaticus*)
17. The Sushil's Bush Frog (*Raorchestes sushili*)
18. The Shillong Bubble-nest Frog (*Raorchestes shillongensis*)
19. The Tiger toad (*Xanthophryne tigerinus*)

(E) FISHES

1. The Pondicherry Shark (*Carcharhinus hemiodon*)
2. The Ganges Shark (*Glyphis gangeticus*)
3. The Knife-tooth Sawfish (*Anoxypristis cuspidata*)
4. Large-tooth Sawfish (*Pristis microdon*)
5. Long-comb Sawfish or Narrow-snout Sawfish (*Pristis zijsron*)
6. The Blue Whale (*Balaenoptera musculus*)
7. Deccan Labeo (*Labeo potail*)

(F) SPIDERS

1. The Rameshwaram Ornamental or Rameshwaram Parachute Spider (*Poecilotheria hanumavilasumica*)
2. The Gooty Tarantula, Metallic Tarantula or Peacock Tarantula (*Poecilotheria metallica*)

(G) CORALS

1. Fire corals (*Millepora boschmai*)

Why to Protect Endangered Species

Wildlife conservation has been advocated through the years by many government and non-government organizations all over the world. Why is there so much concern about wildlife? Why do people have to conserve them? The reasons why wildlife should be conserved are many but among those that stand out are the following:

1. Wildlife is a Source of Medicine

The American Association for the Advancement of Science noted that more than 40,000 species of plants, animals, fungi and microscopic animals are used in some way to benefit humans. More than a thousands of pharmaceuticals originated from wild plants. Life-saving drugs like aspirin, which is useful in controlling blood pressure, to life-saving medicines like Vincristin that can reduce childhood leukemia deaths, were derived from plants. Many extracts from animals can be used as curative and health-sustaining food supplement. Squalene in sharks, for example, is a remarkable nutrient that has many benefits.

2. Wildlife Enriches Food Production

Bees, bats, birds, among other animals help plants become productive by serving as agents of pollination. Pollination is the process by which pollen is transferred in plants, thereby enabling fertilization and sexual reproduction to produce fruit. While feeding, a bee or a bird unwittingly brings pollen or seeds as they fly to feed on a flower or consume the fruits of a plant, respectively. This gives pollen a free ride to a female flower that will facilitate the fertilization process.

3. Wildlife Serves as Models for Man's Innovations and Inventions

Inventors and engineers modelled some submarines on fish movement. Nematodes were used by scientists to model an electronic robot that can detect land mines. Aeroplanes were modelled after birds. These are only some of the many things that arose out of inspiration from life forms.

4. Wildlife Serves as Food to People

Animals like wild pigs, deer, and a whole lot of coastal and marine life like fishes, crabs, molluscs, shells, among others support human population especially in the highly natural resource-dependent countries of the world. Without wildlife as a source of food, many civilizations will not thrive.

5. Wildlife Serves as Gene Pool for Pest-Resistant Strains

Scientists look for resistant strains in wild plants because wild plants have evolved features resistant to disease. Isolating the important genetic component in wild plants can help improve crop production. Special strains can also be used to reduce the impact of global climate change.

6. Wildlife has Bequest Value

The continued existence of wildlife on the planet, especially endangered ones, has bequest value. This will allow the future generations to also enjoy the experience seeing and discovering the unique and awesome characteristics of wildlife.

7. Wildlife Maintains Ecosystem Stability

Ecosystem includes abiotic factors like temperature, humidity etc. and biotic factors like plants and animals. The ecosystems maintain a delicate balance of nature. Each species interacts with other species and plays a role in the transfer of energy and materials within and between ecosystems, hence each one, in its own way, contributes to the stability of ecosystems. The plants occupy the base of food webs, so extinction of a single plant species may lead to the extinction of animal species dependent on that particular plant species.

There may be other undiscovered benefits of utmost importance to man. Wildlife, therefore, must be conserved as much as possible to ensure that the next generation will also enjoy the benefits associated with these creatures.

Widlife Conservation Efforts in India

In the last few decades, human encroachment was one of the biggest threats to facing India's wildlife. Indian Government built many National Parks in order to facilitate endangered species up to some extent. The first park was established in 1935; besides, the Project Tiger was also enacted in 1972. At the beginning of the 20th century, there were around 40,000 tigers in India as compared to the tiger consensus of 2008 which revealed only 1,411 tigers remaining. The 'Project Tiger' is considered to be the most successful conservation project all throughout India. Currently, there are around 39 Project Tiger wildlife reserves in India encompassing an area of more than 37,761 sq.km. Many rhino's are survived and conserved in the Kaziranga National Park. Government makes efforts to conserve the wildlife with the help of either conservation centres or through different agencies, which are as follows

1. Conservation Centers

A number of Conservation Centers have been set up in India for the purpose of studying, propagating, conserving and for the betterment of the highly endangered species of wildlife, both flora and fauna. These Conservation Centers can be divided into Wildlife Reserves, Conservation Centers and Wildlife Sanctuaries.

I. Famous Wildlife Reserves in India

- Sariska Tiger Reserve, Rajasthan
- Bandipur Conservation Center, Karnataka
- Bandhavgarh Conservation Center, Madhya Pradesh
- Corbett Conservation Center, U.P
- Kanha Tiger Reserve, Madhya Pradesh
- Ranthambore Tiger Reserve, Rajasthan

- ☆ Manas Tiger Reserve, Assam
- ☆ Tadoba Tiger Reserve
- ☆ Pench Tiger Reserve
- ☆ Namdapha Tiger Reserve, Arunachal Pradesh

II. Other Wildlife Reserves in India

- ☆ Dudhwa Conservation Center, U.P
- ☆ Srisailam Sanctuary, Andhra Pradesh
- ☆ Pabitora Wildlife Sanctuary, Assam
- ☆ The Palamu Tiger Reserve, Bihar
- ☆ Chandka Elephant Reserve, Orissa
- ☆ Similipal Tiger Reserve, Orissa
- ☆ Palamau Tiger Reserve, Bihar

III. Conservation Centers in India

- ☆ Kaziranga Conservation Center, Assam
- ☆ Royal Chitwan Park
- ☆ Royal Bardia Park
- ☆ Sultanpur Conservation Center
- ☆ Sundarbans
- ☆ Sanjay Gandhi Conservation Center Borivli, Mumbai
- ☆ Rajaji Conservation Center
- ☆ Bharatpur Conservation Center
- ☆ Namdapha Conservation Center, Arunachal Pradesh
- ☆ Dudhwa Conservation Center, Uttar Pradesh
- ☆ Dachigam Conservation Center, Jammu and Kashmir
- ☆ Periyar Conservation Center, Kerala
- ☆ Bannerghata Conservation Center, Karnataka
- ☆ Keibul Lam Jao Conservation Center, Manipur
- ☆ Nagarhole Conservation Center, Karnataka
- ☆ Nandankanan Biological Park, Orissa
- ☆ Valley of Flowers Conservation Center, Uttar Pradesh
- ☆ Nanda Devi Conservation Center, Uttar Pradesh

IV. Famous Wildlife Sanctuaries in India

- ☆ Bharatpur Bird Sanctuary
- ☆ Periyar Wildlife Sanctuary
- ☆ Gir Conservation Center and Sanctuary, Gujarat
- ☆ Kalsubai Harischandragad Sanctuary in the Western Ghats

- ☆ Thattekad Bird Sanctuary in Kerala.
- ☆ Mudumalai Sanctuary, Tamil Nadu
- ☆ Chilika Lake, Orissa
- ☆ Bhitarkanika Sanctuary, Orissa
- ☆ Satkosia - Baisipalli Sanctuary, Orissa
- ☆ Hazaribagh Wildlife Sanctuary, Bihar
- ☆ Lawalong Wildlife Sanctuary, Bihar
- ☆ Dalma Wildlife Sanctuary, Bihar
- ☆ Rajgir Wildlife Sanctuary, Bihar
- ☆ Topchanchi Wildlife Sanctuary, Bihar
- ☆ Kabar Lake Bird Sanctuary, Bihar
- ☆ Kedarnath Sanctuary, Uttar Pradesh
- ☆ Askot Sanctuary, Uttar Pradesh
- ☆ Govind Wildlife Sanctuary, Uttar Pradesh
- ☆ Hastinapur Sanctuary, Uttar Pradesh

2. Conservation Agencies

- ☆ Ministry of Environment and Forests (MoEF): It is the nodal agency in the administrative structure of the Government of India for the planning, promotion, coordination and overseeing the implementation of environmental and forestry programmes. The Ministry is also the Nodal agency in the country for the United Nations Environment Programme (UNEP). The principal activities undertaken by Ministry of Environment and Forests, consist of conservation and survey of flora, fauna, forests and wildlife, prevention and control of pollution, afforestation and regeneration of degraded areas and protection of environment, in the frame work of legislations.
- ☆ Indian Board For Wildlife (IBWL): It is the apex advisory body in the field of Wildlife Conservation in the country and is headed by the Honorable Prime Minister of India. Wildlife and forests shall be declared priority sector at the national level
- ☆ Central Zoo Authority of India (CZA): In India, functioning of zoos is regulated by an autonomous statutory body called Central Zoo Authority which has been constituted under the Wildlife (Protection) Act. The main objective of the authority is to complement the national effort in conservation of wildlife. Standards and norms for housing, upkeep, health care and overall management of animals in zoos has been laid down under the Recognition of Zoo Rules, 1992.
- ☆ Wildlife Institute of India (WII): It is a government institution run by the Indian Council of Forestry Research and Education which trains wildlife managers and wildlife researchers. Trained personnel from WII have contributed in studying and protecting wildlife in India. WII

has also popularized wildlife studies and careers. The institute is based in Dehradun, India. It is located in Chandrabani, which is close to the southern forests of Dehradun. The mission of the Wildlife Institute of India is to train managers and biologists for protected area management and wildlife research

- Zoo Outreach Organization (ZOO): This started off as an NGO primarily focusing on training Zoo staff and bettering the circumstances of captive animals in Indian zoos. It has since evolved into an overall nature and wildlife conservation NGO. Zoo Outreach Organisation (ZOO) was so named because its initial activity involved primarily zoos, and its purpose was to "reach out" and involve more people with zoos.
- Bombay Natural History Society: It is today the largest non-government organization (NGO) in the Indian sub-continent engaged in nature conservation research. Since its establishment in 1883, its commitment has been, and continues to be, the conservation of India's natural wealth, protection of the environment and sustainable use of natural resources for a balanced and healthy development for future generations. The Society's guiding principle has always been that conservation must be based on scientific research
- Wildlife Protection Society of India (WPSI): It was founded in 1994 by Belinda Wright, its Executive Director, who was an award-winning wildlife photographer and film-maker till she took up the cause of conservation. From its inception, WPSI's main aim has been to bring a new focus to the daunting task of tackling India's growing wildlife crisis. It does this by providing support and information to government authorities to combat poaching and the escalating illegal wildlife trade – particularly in wild tigers.
- Wildlife Trust of India (WTI): It is a non-profit conservation organization based in New Delhi. Its principal concerns are crisis management to prevent destruction of India's wildlife and the provision of quick, efficient aid to those areas that require it the most. In the longer term it hopes to achieve, through proactive reforms, an atmosphere conducive to conserving India's wildlife and its habitat. WTI was formed in November 1998 in response to the rapidly deteriorating condition of wildlife. WTI is a registered charity in India
- World Conservation Union or International Union for the Conservation of Nature and Natural Resources (IUCN): This is an international organization dedicated to natural resource conservation. Founded in 1948, its headquarters is located in the Lake Geneva area in Gland, Switzerland. The IUCN brings together 83 states, 108 government agencies, 766 NGOs and 81 international organizations and about 10,000 experts and scientists from countries around the world. IUCN's mission is to influence, encourage and assist societies throughout the world to conserve the integrity and diversity of nature and to ensure that any use of natural resources is equitable and ecologically sustainable. The IUCN

Red List of Threatened Species (also known as the IUCN Red List or Red Data List), created in 1963, is the world's most comprehensive inventory of the global conservation status of plant and animal species.

- Trade Record Analysis of Flora and Fauna in Commerce (TRAFFIC) International: This is an international network for monitoring wildlife trade. Established in 1976 as a joint programme of World Wide Fund for Nature (WWF) and World Conservation Union (IUCN). Based in Cambridge, Great Britain and having offices in Africa, Asia, the Americas, Europe and Oceania. The Organization is sponsored by charitable foundations, private persons and other donors. TRAFFIC's vision is of a world in which trade in wild plants and animals will be managed at sustainable levels without damaging the integrity of ecological systems and in such a manner that it makes a significant contribution to human needs, supports local and national economies and helps to motivate commitments to the conservation of wild species and their habitats.
- Convention on International Trade in Endangered Species of Wild Fauna and Flora (CITES): This is an international agreement between governments, drafted as a result of a resolution adopted in 1963 at a meeting of members of the World Conservation Union (IUCN). Its aim is to ensure that international trade in specimens of wild animals and plants does not threaten their survival and it accords varying degrees of protection to more than 33,000 species of animals and plants. Roughly 5,000 species of animals and 28,000 species of plants are protected by CITES against over-exploitation through international trade.
- World Wide Fund for Nature (WWF): It is an international non-governmental organization for the conservation, research and restoration of the natural environment, formerly named the World Wildlife Fund. It is the world's largest independent conservation organization with over 5 million supporters worldwide, working in more than 90 countries, supporting 15,000 conservation and environmental projects around the world.
- United Nations Convention on Law of the Sea (UNCLOS): Also called the Law of the Sea Convention, it is the international agreement that resulted from the third United Nations Convention on the Law of the Sea that took place from 1973 through 1982 with modifications that were made by the 1994 Agreement on Implementation. The Convention defines the rights and responsibilities of nations in their use of the seas, establishes clear guidelines for businesses, protects the environment, and improves the management of marine natural resources.
- UNESCO's Man and the Biosphere Programme (MAB): The Man and the Biosphere Programme (MAB), proposes an interdisciplinary research agenda and capacity building aiming to improve the relationship of people with their environment globally. Launched in the early 1970s, it notably targets the ecological, social and economic dimensions of biodiversity loss and the reduction of this loss.

Conclusion

Conservation of wildlife wealth in india is a priority. Government and other agencies are making concerted efforts to promote awareness on wildlife, their importance to society, negative impact of poaching and the effect of unmindful human activities on wildlife. Sustained efforts are needed to protect and conserve wildlife to handover them to future generation

References

Bishnoi, S.K. 1988. The Bishnoi movement for saving the blackbuck in Abohar area. Pages 40–43 in A.S. Atwal, S.S. Bains, and M.S. Dhindsa, editors. Wildlife of Punjab. Indian Ecological Society, Ludhiana, India.

Charles Michael Hogan. 2009. Global Twitcher.com ed. N. Stromberg New Zealand Robin: Petroica australis.

Charles Darwin. 1859. The Origin of Species. Reprinted by Doubleday Books, Garden City, New Jersey, USA.

Elliot O. Wilson. 2005. The Future of Life. Alfred A. Knopf. New York, New York, USA.

Erik Flugel. 2004. Microfacies of carbonate rocks: analysis, interpretation and application. Springer Pubishers. 976 pages.

Gilson S. Jamieson and Alan Campbell. 1998. Proceedings of the North Pacific Symposium on Invertebrate Stock Assessment. National Research Council Canada. 462 pages.

Hosette, B. B., and M. Venkateshwarlu. 2001. Trends in wildlife biology conservation and management. Vols. 1 and 2. Daya Publishing House, New Delhi, India.

Lamba, B.S. 1988. Status of Wildlife in Punjab. Pages 20–24 in A.S. Atwal, S.S. Bains, and M.S. Dhindsa, editors. Wildlife of Punjab. Indian Ecological Society, Ludhiana, India.

Laura F. Landweber and Andrew P. Dobson (eds.) 1999. Genetics and the Extinction of Species. Princeton University Press. Princeton, New Jersey, USA.

Madhusudan, M. D. and Mishra, C. (2003). Why big, fierce animals are threatened: conserving large mammals in densely populated landscapes. In Battles over natur: science and the politics of conservation: 31–55, Saberwal, V. and Rangarajan, M. (Eds) New Delhi: Permanent Black.

Oppa L.N., Loarie S. R., Pimm, S.l. (2008). On the Protectionof Protected Areas. Proceedings of the National Academy of Sciences 105: 6673–6678.

Pitt Nystrom. 1999. Ecological impact of introduced and native crayfish on freshwater communities: European perspectives. In Gherardi, F. and Holdich, D.M. (eds.) Crustacean Issues 11: Crayfish in Europe as Alien Species (How to make the best of a bad situation?) A.A. Balkema, Rotterdam, Netherlands: 63–85.

Parshad, V.R. 1988. Mammalian fauna of Punjab. Pages 25–38 in A.S. Atwal, S.S. Bains, and M.S. Dhindsa, editors. Wildlife of Punjab. Indian Ecological Society, Ludhiana, India.

Paul R. Ehrlich and Anne H. Ehrlich. 1981. Extinction: The causes and consequences of the disappearance of species. Random House, New York, New York, USA.

Proc. Species at Risk 2004. Pathways to Recovery Conference. 11 March 2–6, 2004, Victoria, B.C.

Rahmani, A.R. 2002. The deteriorating status of the Indian bustard. *Oriental Bird Club Bulletin* 35: 45.

Rangarajan M (2001) India's Wildlife History, Permanent Black New Delhi.

Rodgers, W. A., Panwar, H. S. 1988. Planning a protected area network in India. Dehradun, Wildlife Institute of India.

Rodgers, W. A., Panwar, H. S. and Mathur V. B. 2000. Wildlife protected area network in India. – A review. Dehradun, Wildlife Institute of India.

Sekhar, N. (2003). Local people's attitudes towards conservation and wildlife tourism around Sariska Tiger Reserve, India. *Journal of Environmental Managemnt* 69: 339–347.

Shahabudin G, Rangarajan, M. (2007). Introduction. In: Shahabuddin G, Ranagarajan, M. (eds) Making conservation work. Permanent Black, New Delhi

Tolly Dc Hooper, editor. Proceedings of the Species at Risk. 2004. Pathways to Recovery Conference. 1 March 2–6, 2004, Victoria, B.C. Species at Risk 2004 Pathways to Recovery Conference Organizing Committee, Victoria, B.C..

Chapter 5

Conservation and Management of Indian Gangetic Dolphin

Seema Jain[1] and Manu Varma

Department of Zoology
R.G.P.G. College, Meerut, U.P.
[1]E-mail: seema61@gmail.com

ABSTRACT

The Indian Gangetic dolphin, Platanista gangetica gangetica is found only in the Indian sub-continent. National Ganga River Basin authority (NGRBA) on October 5, 2009 declared it as ***National Aquatic Animal.*** *Due to various pressures the distributional ranges and abundance of this species declined sharply. At present not more than 2500 individuals of this species exist in its entire distributional range. Therefore, it has been categorised as* ***"endangered"*** *on IUCN Red List. Workshop jointly organized by National Mission for Clean Ganga and WWF – India on March 19–20, 2014 laid down objectives Dolphin Action Plan 2010–2020.*

Keywords: *Gangetic Dolphin, Conservation, Management.*

Introduction

The Ganges River Dolphin, is often known as the 'Tiger of the Ganges', is an indicator animal and has the same position in a river ecosystem as a tiger in a forest, its presence indicating a sign of a healthy river ecosystem It is an endemic and rare charismatic mega-fauna found only in the Indian subcontinent. Gangetic dolphin (*Platanista gangetica gangetica*) is found in the Ganges–Brahmaputra–Meghna and Karnaphuli River systems of India, Nepal and Bangladesh (Anderson, 1878; Kasua, 1972; Jones, 1982; Reeves *et al.*, 1853; Shrestha, 1989; Reeves, 1993). The Prime Minister of India declared the Ganges river dolphin the **National Aquatic Animal** in the First Meeting of the National Ganga River Basin Authority (NGRBA) on October 5, 2009. The Ministry of Environment and Forests (MoEF) subsequently notified that the Ganges river dolphin is the National Aquatic Animal on May 10, 2010 and constituted a working group to prepare a conservation action plan for the preservation of the Ganges river dolphin. Most of the cetacean researchers consider the occurrence of only four species of freshwater river dolphins (Boto, Baiji, Bhulan and Gangetic dolphin) in the world. In the nineteenth century, the dolphins were plentiful in the entire distributional range, though no actual data on population is available (Sinha and Sharma, 2003). However, due to various pressures the distributional ranges and abundance of this species has sharply in its entire distributional ranges (Reeves and Leatherwood 1995). At present there are not more than 2500 not reversible individuals of this species in the world (Sinha and Sharma 2003). Their total population is estimated to be 3500 (in 2014) individuals in its entire distribution range, of which more than 80 per cent is within Indian territory (Sinha *et al.*). Compared to the Ganges river system, studies on dolphin or its conservation are scanty in the Brahmaputra river system. Earlier studies on the Gangetic dolphin were mainly focussed on population status (Mohan 1997, 1998, Biswas 1997, Wakid 2006) ecology (Biswas 2000) and threats (Bairagi 1999). This species has been included in Schedule I of the Indian Wildlife (Protection) Act, 1972, in Appendix I of the Convention on International Trade in Endangered Species (CITES), in Appendix II of the Convention on Migratory Species (CMS), and is categorized as "endangered" on the International Union for the Conservation of Nature's (IUCN) Red List.

Historical Review

During the time of King Ashoka, the Maurya Emperor in 230 B.C., the Gangetic dolphin was known as 'Ganga-Puputaka'. It enjoyed the status of a 'protected animal' through a decree known as 'Fifth Pillar Edict'. In *Baburnama,* which was published in 1598, the Gangetic dolphin was known as *Khokk Aabi* or Water Hog. Bhattasali, the author of *Iconography of Buddhist and Brahminical sculptures* identified this species as the *Bahana* (carrier) of mythical goddess Ganga.

The Ganges Dolphin was first described by William Roxburgh and Heinrich Julius Lebeck in 1801. Priority for the description goes to Dr. Roxburgh who named it *Delphinus gangetica* (Pilleri 1978). Lebeck published his work on Gangetic dolphin in 1801 in Germany. The name *Susu platanista* was given by Rene Lesson in 1828.

However, Johann Wagler adopted *Platanista* as a genus in 1930 and from then onwards the Gangetic dolphin is known as *Platanista gangetica* to the scientific community. The Indus and Ganges populations were regarded as identical. Pilleri and Gihr (1971) divided Ganga river dolphin and Indus dolphin into two species based on differences in skull structure.

One species is recognized in the genus *Platanista;* the Ganges River dolphins are currently *Platanista gangetica gangetica* and the Indus River dolphins are *Platanista gangetica minor.*

The present species is commonly known as *Susu* in India, *Shunshuk* in Bangladesh and *Saunch* or *Sounch* in Nepal.

The species has different local names, *i.e.*, in India this species has different names in different parts, *viz.*, *Susuk* (West Bengal), *Susu* (Bihar), *Hihu* (Assamese), *Putukari* (in Mishing).

Taxonomic Status

Phylum:	Chordata
Sub-phylum:	Vertebrata
Class:	Mammalia
Order:	Cetacea
Suborder:	Odontoceti
Family:	Platanistidae
Genus:	*Platanista*
Species:	*gangetica*
Subspecies:	*gangetica*

Primitive Characters

Platanista gangetica gangetica bears some very primitive characters not known in other cetaceans, such as the presence of a caecum at the junction of the small and large intestines. Sinha *et al.* (2010) suggest that the position of the testis is more dorsal than that in other marine cetaceans and subcutaneous muscle is present between two layers of blubber.

External Morphology

Platanista gangetica gangetica have a long, pointed snout that is the characteristic of all river dolphins. In female's snout is generally longer than that of the male and may curve upwards and to one side. Both the upper and lower jaw have sets of long sharp teeth. The eyes are extremely small openings slightly above the mouth. The body is deep brown in colour, stocky in the middle and attenuating to a narrow tail stalk behind the dorsal fin. The dorsal fin is somewhat triangular hump located two-thirds of the body length from the anterior end. The flippers are broad and have a crenellated margin, with visible hand and arm bones. The flippers and flukes are thin and large in relation to the body size. Body length is about 2–2.2 m in males and 2.4–2.6 m in females. The young ones measure 70–90 cm and weigh between 4–7.5 kg at the time of birth. While adults usually weigh between 70–90 kg. An adult pregnant female (2.5 m) can weigh up to 114 kg.

Behaviour

Behara and Rao (1999) observed some interesting behaviours of Gangetic dolphin. Verma *et al.* (2004) investigated the phylogenetic relationship of the species with other cetaceans and found that the species is closer to Mysticeti than any other toothed whales.

Ganges river Dolphin prefers to stay in deep waters in and around the confluence of two or more rivers and shares its habitat with crocodiles, freshwater turtles and wetland birds many of which are fish-eaters and are potential competitors with dolphins. It exhibits greater preference for the surface than other river dolphins; even when swimming, which they do on their sides, they occasionally keep their beak out of water. It can swim and vocalize constantly. The underwater activities of Ganges dolphins in the wild are difficult to observe due to turbidity of the river. Their short surfacing time is also a major constraint for behavioural studies. A recent study recorded six types of surfacing patterns, which were dependent on age-class and off-shore distance of the individual (Sinha *et al.*, 2010a). Ganges newborn calves frequently leap completely out of the water.

Food and Feeding Habit

The dolphins have been observed chasing and preying upon surface dwelling fish species, such as *Rhinomugil corsula*. These are catholic feeders and feed on several species of fishes, invertebrates (Sinha *et al.*, 1993). They exhibit active foraging behaviour in the morning (0700–1000 hrs) and afternoon (1500–1700 hrs). Several studies revealed that they drive fishes to a particular area for community feeding.

Migration and Dispersal

During October to April, many dolphins leave the tributaries of the Ganga and Brahmaputra systems and congregate in the main channels, only to return to the tributaries the following monsoon. Dolphins were seen more than 150 km upstream in the River Son in the flood season in the early 1990s, returning to the main stream of the Ganga after the floods abated in October. According to Reeves and Brownell (1989) they may become isolated in pools and river branches during the dry season.

Life History

Mating usually takes place during March to June, it has been observed even in July. The breeding season of the Ganges dolphin extends from January to June, newly-born calves can be seen even in other months. Only a single baby is born after a gestation period of about 9 months. At the time of birth the neonate is about 70–90 cm and weighs about 4–7.5 kg. The mother and calf remain together for about one year. According to Kasuya (1972) the male attains sexual maturity at an age of about 10 years when they reach a length of 1.7 meters while the females are known to attain sexual maturity at 10 or less years, when they are around 2m long (Harison, 1972).

Distribution

There is no conclusive fossil record, still it is estimated that the Gangetic dolphin evolved around 20 million years ago. After Roxburgh (1801) and Lebeck (1801), Anderson (1878) gave a detail distribution pattern of the species in India, where he mentioned the distribution of the species from Haridwar to Sundarban delta region. After that Eschrict (1852), Jerdon (1874), Sterndale (1884) and Blanford (1891) reported especially on the distribution, behaviour and anatomy of the species.

After the classical works of Anderson (1878) on the dolphins of Bangladesh, Jones (1982) recorded the maximum upstream distribution of dolphins in Meghna river system of the country. Jones (1974) suggested the introduction of Gangetic dolphin in some new habitats of Indian subcontinent for their better protection. Nath (1974) published a report on the habit and habitat of dolphin in Ganga near Patna. Haque (1976) reported the abundance and distribution and the effect of Farrakka barrage on the dolphin population. Again, Haque *et al.* (1977) described the biology and behaviour of Gangetic dolphin. Gupta (1986) reported the distribution pattern of dolphin in certain segments of Ganga. Gupta (1986) described his 27 surveys in scattered locations of the Ganges between Allahabad and Calcutta in 1978; Ali (1992) conducted surveys from Buxar to Sirighat, a stretch of Ganges of 463 km on 6-14th August 1989. Sinha (1997) conducted another survey in the Bhagirathi River from the Janjipur Barrage to Tribenighat (approx. 320 km). Sinha (1997) also conducted another survey in the Hoogly River from Tribenighat to Calcutta Botanical Garden (approx. 100 km). Due to very low population status and rapidly declining distribution ranges, the action plan of IUCN/SSC Cetacean Specialist Group calls attention to the need for range-wide population assessments in support of conservation effort for the Gangetic dolphin as well as its other three close relatives of the world (Reeves and Leatherwood 1995, Reeves *et al.*, 2000).

Population Status

Without a standard survey-design it is impossible to detect trends in absolute population abundance (Reeves *et al.*, 1993). Besides, Reeves and Brownell (1989), Reeves *et al.* (1993), Ahmed (2000) and Smith *et al.* (2001) reported the status of dolphins in the Karnaphuli River, Halda River, Matamuhuri River, Bangkhali River and Kaptai Lake of the country. Choudhury and Hussain (1992) reviewed the population status of dolphin in the protected areas of India and also suggested protection measures for this species. In Chambal River, Singh and Sharma (1985), Rao *et al.* (1989), Sharma (1993) and Sharma *et al.* (1995) investigated the population status at different times. Rao (1995) investigated the population status of dolphins in the Ganges main stream in between Bijnor and Narora Barrages (approx. 166 km). Smith *et al.* (1998) described the effects of water development projects on dolphin population in Bangladesh. Singh and Sharma (1985) estimated the population status of dolphin in Chambal River, a south-western tributary of Ganga. Rao *et al.* (1989) and Hussain and Choudhury (1992) documented the population status and ecological status of Gangetic dolphin in the National Chambal Sanctuary. In Nepal, Shrestha (1989 and 1995), Smith (1993), Smith *et al.* (1994) and Sinha *et al.* (2000) investigated the population status and distribution pattern of dolphins in the major rivers *viz.*, Karnali, Narayani, Mahakali and Kosi Rivers.

Recently, Wakid (2005) assessed the population status and distribution pattern of dolphin population in and around of Dibru-Saikhowa National Park of Eastern Assam and Singha *et al*2013, assessed river Barak of Assam. Braulik *et al.*, 2014, assessed the population of this species in River Kosi and son of Bihar. The marked seasonal changes in the dolphin distribution and density over much of its range are due, at least in large part, to fluctuations in water levels.

Conservation and Management

Threats

The "Tiger of the Ganges" or the Ganges River Dolphins are facing increasing threats to their environment and are on the verge extinction, due to tremendous anthropogenic pressures in their distributed ranges. It is estimated that not more than two thousand Ganges river Dolphins are alive today in its entire distribution range. In the Ganga basin, the river dolphins have declined in abundance and in the extent of their range. The present species is facing several threats and verge are on of extension due to following reasons:

Direct Catch or Poaching

Deliberate killing of Dolphin, especially through harpooning, for their oil and meat, used as liniment and as an aphrodisiac.

Incidental Catch or Accidental Killing

Accidental killing is a severe problem for Ganges River dolphin throughout most of their range. The primary cause is believed to be entanglement in fishing gear, most often nylon gillnets because their preferred habitat is often in the same location as primary fishing grounds.

Habitat Fragmentation and Habitat Degradation

Development of hydroelectric power and irrigation in the Ganges River system has prevented dolphin migration and has led to the segregation of populations and consequently leading to a narrowed gene pool. These developments have also altered the overflow and flooding cycles of the Ganga river system, which influences the movements and spawning of flood-plain dependent fishes. Other sources of habitat degradation include the removal of stones, sand (Mohan *et al.*, 1997) and woody debris (Smith, 1993). These activities threaten the ecological integrity of the riverine environment, especially in small tributaries where suitable habitat is more confined and therefore more vulnerable to local sources of degradation.

Construction of Dams and Barrages

In addition to fragmenting dolphin populations, dams and barrages degrade downstream habitat and create reservoirs with high sedimentation and altered assemblages of fish and invertebrates (IWC, 2000). Luxuriant growth of macrophytes and excessive siltation have eliminated suitable habitat immediately above Farakka Barrage (Sinha, 2000). Riverine resource extraction dredging and development of the river environment has altered its nature and eliminated counter currents, one of the most preferred habitats of the dolphin.

Excess Abstraction

The river water for irrigation has lowered water levels throughout species range and has threatened suitable habitat especially in the Ganges where the mean dry-season water depth has been dramatically reduced in the recent years.

Pollution

The concentrations of fertilizer and pesticides residue, and industrial and domestic effluents are very high in the Ganges River. The effects of pollutants may be deleterious to dolphin populations, and pollutant loads are expected to increase with industrialization and the spread of intensive modern agricultural practices (Smith and Reeves 2000a). A recently discovered micro-pollutant, perfluorinated compounds (PFCs), was assessed in tissues of the dolphin carcasses collected from the Ganga in and around Patna (Yeung *et al.*, 2009).

Depletion of Prey Base

Sustained and heavy exploitation of small fishes of rivers by the wide spread use of the mosquito mesh nets particularly in the river Ganga. Further research is required to quantify this threat in larger areas.

Overexploitation

It is also a major cause of the depletion of river dolphin.

Use of Dolphin Products

The fishermen and other riparian people in many areas have traditionally been using dolphin oil for burning and as a liniment (Anderson 1879). Some people like to eat the meat and in Bangladesh pregnant women consume the dolphin oil (Smith *et al.*, 1998). Currently the dolphin oil is also used as fish bait in Bihar and Assam for catching two economically important fish, *Eutropiichthys vacha* and *Clupisoma garua* (Sinha 2002).

Possible Impacts of River Traffic

There have been numerous proposals to utilize sections of the Ganga for inland water transport. Currently, cargo steamers carry goods from Haldia to Patna regularly, occasionally travelling up to Varanasi. There is heavy river traffic in and around Kolkata and river tourism has recently been started between Kolkata and Varanasi. Collisions with these vessels, and the effects of the noise they generate, are a serious concern for the dolphins. Since this species depends upon echolocation for many of its activities, including foraging, sonic pollution may adversely impact their well-being. The Ganges dolphins were seen to be avoiding the river stretch with heavy traffic in the River Hooghly in and around Kolkata, preferring relatively undisturbed areas where a lower number of mechanized vessels ply (Sinha 1997).

Conservation Strategies

Platanista gangetica gangetica has been included in Schedule I of the Indian Wildlife (Protection) Act 1972, in Appendix I of the Convention on International Trade in Endangered Species (CITES), in Appendix II of the Convention on Migratory Species (CMS) and categorized as Endangered by the IUCN (International Union for Conservation of Nature). Dolphin conservation, like any other conservation programme, requires public engagement, awareness generation on the existing Government policies and mobilizing key stakeholders (communities, businesses and government machinery) for field action and research, thereby winning their support for the species conservation and making them active change agents in the form of Mitras (friends) of the River. The programme has been assessing the Environment Flows or E-Flows in Ganga and Ramganga, to determine the required flows in the river to meet its ecological needs. WWF-India has been undertaking several initiatives towards the conservation of Ganges river Dolphin in India while garnering partnerships with various civil societies, Government Departments, academicians, researchers and Universities to continue the work for the species' protection.

One of the key early initiatives to protect the species was taken in 2005, when Bridge Ghat to Narora was declared as a RAMSAR site by the Government of India. Ramsar sites are important aquatic habitats such as wetlands or riparian and coastal zones of international importance, designated under the Ramsar Convention. These habitats are biodiversity rich areas supporting vulnerable, endangered, or critically endangered species or threatened ecological communities.

Another milestone for Dolphin conservation was achieved when National Ganga River Basin Authority (NGRBA) was constituted by the Government of India under the Chairmanship of the Prime Minister on 20th February 2009, in which the Prime Minister declared the Ganges Dolphin as the National Aquatic Animal. The Ministry of Environment and Forests notified the Ganges river Dolphin as the National Aquatic Animal on 18th May 2010. The Ministry also constituted a six-member Working Group under the chairmanship of Prof. R K Sinha on October 28, 2009, to prepare an Action Plan for Conservation of Ganges River Dolphin in the Ganga River, which is considered to be one of the major bio-indicators of ecological health of Ganga.

WWF-India has also been closely working with river bank communities (fishermen, farmers, school children, etc.) and promoting Smart Agricultural Practices to reduce river pollution and over abstraction of river water for irrigation.

Just as the Tiger represents the health of the forest, just as the Snow Leopard represents the health of the mountainous regions, and the Cheetah, the health of grasslands, the presence of the Dolphin in a river system signals its good health and biodiversity. This move is expected to increase support for the Government's efforts to protect this vulnerable species. As an effort to assess the total number of species and bring attention to its depleting numbers, under WWF-India's 'My Ganga, My Dolphin' campaign launched in October 2012, a participatory comprehensive scientific survey, across 2,500 km stretch of Ganga and its tributaries was organized. The survey declared the presence of 671 dolphins in the upper Ganga system. The outcome of the campaign was preparation of a short and a long-term Action Plan for Dolphin conservation, its habitat and the river.

Recently in March 2014, a workshop on the 'Operationalisation of the Dolphin Action Plan' was organized by National Mission for Clean Ganga, Ministry of Environment and Forests and WWF-India. The workshop outlined timelines, deliverables, budget, and the roles and responsibilities of various organizations and departments at state and the Centre to develop an operational plan to implement the Dolphin strategy and a unified methodology for its population census. It was decided to launch a national survey to assess the population of the survey and emphasize community engagement for the conservation of the species and identify community reserves. WWF-India believes it is imperative that with the technical strength possessed by institutions, organizations, experts and academicians is channelized to save the declining population of this species. It is also essential to engage with stakeholders and uniquely position the conservation programme to bring the people, river bank communities (as they are intricately involved in Dolphin habitat depletion), businesses, civil society and policymakers to the table that have the passion, expertise and required resources for the restoration of Dolphin. What is needed is a paradigm shift in the way communities look at rivers in India and the biodiversity they support. Restoring the ecological health of the river will need a multidisciplinary approach and one is required to see river conservation in a holistic way. It involves social, technical, institutional and fiscal strategies to promote policies and practices on the ground. This will also require mobilizing various competing water users to come together to address their shared water risks using a coordinated multi-disciplinary framework. The present data is based on the secondary data collected from different sources like WWF, Freshwater and Wetland Conservation Programme WWF-India (Ganga case study), IUCN (International Union for Conservation of Nature), The conservation action plan for the Gangetic dolphin; 2010-2020, The Ministry of Environment and Forests; India, journals, magazines, newspapers etc. And The main aim of this paper is to assess the current population status, distribution and habitat ecology of endangered Ganges River Dolphin (*Platanista gangetica gangetica*) in Ganga River System of India and to assist the Government, Conservation organization and local community to develop conservation action plan providing them the recent scientific facts about *Platanista gangetica gangetica* in Ganga River System. The secondary aim of this paper is to aware people about the conservation importance of freshwater river dolphin to enhance their active participation in dolphin conservation intervention. The Gangaetic Dolphins have narrow ecological requirements and a fragmented population structure. Conserving this species requires coordinated efforts among agencies, organisations, and communities within the species range. Conservation issues can best be addressed by adopting population or regional level approaches for sustainable co-management.

Successful strategies to facilitate the recovery of depleted populations, reverse trends of population decline and habitat deterioration, and to ensure that robust populations with high-quality habitat are secure will need to be multifaceted, adaptable, and tailored to particular local or regional conditions (Reeves *et al.*, 2003). The many elements outlined below are integral to a comprehensive conservation strategy for the Ganges River dolphin. Ensuring that any catch or other uses of the Ganges dolphin are sustainable. The following factors make the deliberate exploitation of Ganges dolphin a high-risk endeavour from a conservation viewpoint:

(a) They have intrinsically low rates of population increase;

(b) Most populations are subjected to by-catch in fisheries;

(c) There is uncertainty associated with estimates of their life history parameters, absolute abundance, trends in abundance, and total mortality;

(d) The effects of chemical and noise pollution, reduced prey abundance, and habitat degradation are potentially serious but difficult to quantify. Ganges River dolphins are especially vulnerable because of their inland freshwater distribution.

A complicating factor is that their small size makes the carcasses easy to handle, transport, process and conceal from management authorities. Measures to regulate directed takes are not easy to devise and implement, but without them, the species population is at risk.

Among the elements needed for an effective management scheme are abundance estimation and reliable demographic knowledge of how many animals are being removed. Developing and encouraging alternative fishing techniques. The use of dolphin oil and their body parts creates an incentive for hunting dolphins and a disincentive for fishermen to release any that may become entangled in their nets. Oil extracted from fish offal available at outdoor markets has been tested and found to be an effective substitute for dolphin oil (Sinha, 2002), and encouraging its use may result in a reduction of dolphin kills.

Reducing incidental mortality through rescue and release efforts A programme was initiated in Pakistan in 2000 to rescue Indus river dolphins that are stranded in irrigation canals or isolated deep pools where they are unlikely to survive (Braulik, 2000). A similar programme in Indian river systems for the Ganges River dolphin will contribute to its conservation. Managing future dolphin-oriented tourism though no dolphin-oriented tourism exists on the Ganga yet, recently introduced river tourism between Kolkata and Varanasi attracted foreign tourists to watch dolphins in the Ganga. While dolphin watching is a potential sustainable income source for fishermen and locals, it is important to ensure that dolphin watching on the Ganga is conducted in a manner that is respectful to the animals, local people and fellow tourists.

Guidelines and codes of conduct should be developed, adopted and promoted by both the tourism industry as well as by government agencies to ensure that unregulated dolphin watch tourism does not become a menace for the Ganges River dolphin. Guidelines were based on:

Habitat Protection and Restoration

Dolphin monitoring units should be formed in association with local communities and management authorities in the identified important dolphin habitats. The units should be encouraged to closely monitor the dolphins and their habitats. All these units need to work together as a single dolphin conservation network through information dissemination and simultaneous actions.

During the rainy season, dolphins usually migrate through the tributaries of Brahmaputra River. Steps should be taken to protect these seasonally migrating dolphins. All the tributary mouths are treated as important dolphin habitats and fishing controlled in these tributary mouths. Close monitoring of these river mouths, especially during rainy season, should be made with the help of a local dolphin monitoring group.

Poaching area and poachers should be identified and strong legal actions should be taken against poaching.

Protected Areas

Existing protected areas in the Ganga and its tributaries fall far short of what is needed. A major challenge in extending the coverage and level of protection conferred through protected areas is to convince stakeholders that conservation measures can benefit them and thus deserve their support. Community Reserve may be an alternative choice. While a buffer zone aids in protection, animals that are relatively safe from entanglement in fishing gear while inside a reserve with strong enforcement may meet a gauntlet of nets as they move beyond its borders.

Researching and Reducing Environmental Pollution

While more research is needed to elucidate the impacts of contaminant exposure on dolphin health, the precautionary principle demands that restrictions on the discharge of untreated effluents into the Ganga river system should be enforced to ensure a reduction in dolphin exposure to toxins.

Reducing the Effects of Water Development on Rivers in Ganga Basin

The Ganges River dolphins need to be considered in the assessment of impacts of water development projects. The preferred option from a conservation perspective is to refrain from interfering with the natural flow regime and to avoid constructing barriers to animals and sediment movement. However, socio-political conditions make it impractical to completely halt water development activities especially in the Ganga basin, so the immediate goal must be to manage such activities in ways that will minimize the harm to dolphins and other aquatic species.

Enhancing the Capacity and Governance Framework for Ganges Dolphin Conservation

A detailed study should be undertaken on the dolphin by-catch mortality. In areas where this problem is prevalent, proper identification of the fishing gears and concerned communities are necessary. Steps should be undertaken to modify the identified fishing gears and practices, the State Fishery Department to undertake legal actions to ensure the control of such catch by these fishing gears, and local NGOs to create awareness in the identified communities about the problem.

A detailed scientific study should be undertaken on the ecology, behaviour, biology and genetics of the Gangetic dolphin, which will help in the long-term conservation of the species in the Ganga and Brahmaputra river system.

Hightlights of Dolphin Action Plan 2010-2020

A study to identify high-density dolphin areas and hot-spots to be protected from illegal fishing and commercial sand mining, and a review of inland fisheries policy and its implementation at the State level has also been recommended. Overall Matrix for Operationalization of Dolphin Action Plan, 2010–2020.

- Initiating state-wise Ganges River Dolphin population status surveys and threat assessment: October to November, 2014
- Setting up of Protected Areas for the Ganges River Dolphin: (March 2015)
- Education and Awareness Immediate Capacity building: February, 2015
- Emerging areas of research: Immediate
- Critical level of flow in riverine habitats, commercial fishing: June, 2015
- Sand mining June, 2015 Initiating state-wise Ganges dolphin population status surveys and threat assessment:Every year (October to November)
- Setting up of Protected Areas for the Ganges river dolphin: 2016
- Education and Awareness: 2014 to 2017
- Critical level of flow in riverine habitats, Commercial fishing: 2015 to 2019
- Sand mining: 2016.

References

Ahmed, B. (2000): Water development and the status of the shushuk (*Platanista Gangetica*) in Southeast Bangladesh. In *Biology and Conservation of Freshwater Cetaceans in Asia*, Occasional Papers of the IUCN Species Survival Commission no. 23 (eds R. R. Reeves, B. D. Smith and T. Kasuya), pp. 62–66. IUCN, Gland, Switzerland.

Ali, S. M. (1992): The Gangetic dolphin. *My Forest*. 28(3): 24–25.

Anderson, J. (1878): Anatomical and zoological researches comprising an account of zoological results of two expeditions of western Yunnan in 1868 and 1875, and a monograph of two cetacean genera, *Platanista* and *Orcella*. B. Quatrich, London, 2vol.

Bairagi, S.P. (1999): Oil bait fishery of catfishes in Brahmaputra River affecting river dolphin populations in Assam, India. *Journal of the Bombay Natural History Societ* 96: 424–426.

Bairagi, S.P. (2003): River dolphins in Brahmaputra system, threats from water development projects. *The Ecologist Asia*. Vol. 11 (1): 28.

Behera, S. and R. J. Rao (1999): Observations on the behaviour of Gangetic dolphins *Platanista gangetica* in the upper Ganga river. *Bombay nat. Hist. Soc.* 96(1): 42-47.

Biswas, S. P. and S. Baruah (2000): Ecology of river dolphin (*Platanista gangetica*) in the Upper Brahmaputra. *Hydrobiologia*, 430: 97–111.

Biswas, S. P., A. Baruah and R. S. L. Mohan, (1997): Current status of river dolphin (*Platanista gangetica*) in the river Brahmaputra. *Internatl. J. Ecol. Environ. Sci.*, 23: 357–361.

Blanford, W. T. (1891): The fauna of British India, including Ceylon and Burma. *Mammalia*. Taylor and Francis, London, Part II, pp. 251–617.

Braulik, G.T. Arshad Masood, Uzma Noureen, Simon.P.Northridge (2014): Current Status of the Ganges River Dolphin, ***PLATANISTA GANGETICA*** in the Rivers

Kosi and Son, Bihar, India. 1PLOS ONE | www.plosone.org Volume 9, Issue 7, e101657.

Choudhury, A. U. (1997): *Checklist of the Mammals of Assam*. Gibbons Books, Guwahati. 103 pp.

Choudhary, B. C. and S. A. Hussain (1992): Ganges river dolphin in protected areas of India: Review of it's status and conservation needs. *Proc. Seminar on Conservation of River dolphin of the Indian Sub-Continent*, 18–19 august, New Delhi. 76

Eschrist, D. F. (1852): On the gangetic dolphin. *Ann Mag. nat. Hist* (2) 9: 160–279.

Gupta, P. D. (1986): The gangetic dolphin *Platanista gangetica* (Lebeck, 1801) pp. 553–562. *In* T.C. Majupuria (ed) *Wildlife Wealth of India* (Resource and Management), Teepress Service, LP. Bangkok.

Haque, A. K. M. Aminul (1976): Comments on the abundance and distribution of the Ganges susu, *Platanista gangetica* and the effect of the Farkka Barrage on its population. *Proc. Scientific Consultation on the Conservation and Management of Marine Mammals and their Environment*, Bager, Norway, 31 August – 9 September 1976 ACMMR/MM/SC/32 pp.

Haque, A. K. M. Aminul, M. Niswaki, M. Kasuya and T. Tobayama (1977): Observation on the behaviour and other biological aspects of the Ganges susu, *Platanista gangetica. Scientific Reports of Whales Research Institute* (*Tokyo*), 29: 87–94.

Harrison, R. J. (1972): Reproduction and reproductive organs in *Platanista indi* and *Pltanista gangetica. Invest. Cetacea*, 4: 71–82.

Hussain, S. A. and B. C. Choudhury (1992): The Gangetic River Dolphin (*Platanista gangetica*) and the status of its habitat in National Chambal Sanctuary. *Proc. Seminar on Conservation of River dolphin of the Indian Sub-Continent*, 18–19 August, New Delhi.

IUCN (1996): *1996 IUCN Red List of Threatened Animals*. IUCN, Gland, Switzerland and Cambridge, UK. 448pp.

Jerdon, T. C. (1874): *The Mammals of India, Cetaceans, a natural History of all animals known to inhibit continental India.* pp. 155–162. John Wheldon, London.

Jones, S. A. (1974): A suggestion for the introduction of the Indus susu and the Gangetic susu into the new sectors and river systems in the sub-continent and establishment of sanctuaries for them. *Proc. Scientific Consultation on the Conservation and Management of the Marine Mammals and their Environment.* Bergen, Norway. ACMRR/SC/16–23 pp.

Jones, S. (1982): The present status of the Gangetic susu, *Platanista gangetica* (Roxburgh), with comments on the Indus susu, *P. minor* (Owen). FAO Advisory Committee on Marine Resources Research, Working Party on Marine Mammals. *FAO Fisheries Series*, 5(4): 97–115.

Kasuya, T. (1972): Some comments on the growth of Ganges dolphin with a comment on the Indus dolphin, *Sci. Rep. Whales Res. Inst.*, 24: 109–115.

Klinowska, M. (1991): *Dolphins, Porpoises and Whales of the World*. The IUCN Cetacean Red Data Book. IUCN, Gland, Switzerland and Cambridge, UK, 429 pp.

Kukenthal, W. (1909): *Untersuchungen an Walen*. Jenaischen Zeitschrift fur Naturwissenschaft. 45: 545–588.

Kumari A, R. K. Sinha, K. Gopal and S. Lata (2002): Concentration of organochlorines in Ganges river dolphins from Patna, Bihar. *J. Environ. Biol.*, 23(3): 279–281.

Lebeck, H. J. (1801): *Delphinus gangeticus* beschrieben von H. J. Lebeck, zu Trankenbar. *Neue Schr. Ges. Natur. Freunde*, 3: 280–282.

Liu, R. and D. Wing (2000): Effects of water development on river cetaceans in China. In: *Biology and Conservation of Freshwater Cetaceans in Asia*, Occasional Papers of the IUCN Species Survival Commission no. 23 (eds R. R. Reeves, B. D. Smith and T. Kasuya), pp. 40–42. IUCN, Gland, Switzerland.

Mohan, R. S. L. (1989): Conservation and management of Ganges River Dolphin *Platanista gangetica* in India. pp. 64–69. *In*: W.F. Perrin, R.L. Brownell Jr., Zhou kaiya and Liu Jiankang (eds.) *Biology and Conservation of the River Dolphins*. IUCN Species Survival Commission Occasional Paper No. 3. IUCN, Gland, Switzerland.

Mohan, R. S. L., S. C. Dey, S. P. Bairagi and S. Roy (1997): On a survey of the Ganges River Dolphin, *Platanista gangetica* of the Brahmaputra river, Assam. *J.Bombay nat Hist. Soc.*, 94(3): 483 - 495.

Mohan, R. S. L., S. C. Dey and S. P. Bairagi (1998): On a residential dolphin population of the Ganges river dolphin, *Platanista gangetica* in the Kulsi river (Assam), a tributary of Brahmaputra. *Ibid*, 95(1): 1–7.

Nath, B. (1974): On some aspects of habit and habitat of the Gangetic dolphin (*Platanista gangetica*)(Lebeck) in the river Ganges at Patna. *Naturalist. Bull. Bihar Nat. Hist. Soc.*, 1(2): 6–7.

Norman, J. R. and F. C. Fraser (1948): *Giant fishes, whales and dolphins*. Putman Press, Bath, England. 376 pp.

Pilleri, G. (1970): Observations on the behaviour of *Platanista gangetica* in the Indus and Brahmaputra rivers. *Invest. Cetacea*, 2: 27–60.

Pilleri, G. and M. Gihr, (1971): Difference observed in the skull of *Platanista gangetica* and *Platanista indi* Blyth, 1859. *Invest. Cetacea*, 3(10): 13–21.

Pilleri, G. (1978), William Roxburgh (1751-1851), Heinrich Julius Lebeck (1801) and the discovery of the Gangetic Dolphin (*Platanista Gangetica*, Roxburgh,1801) Invest. Cetacea, 9: 11–21.

Rao, R. J. (1995): Studies on biological restoration of Ganga River in Uttar Pradesh: an indicator species approach. Final technical report submitted to the Ganga Project Directorate, Government of India.

Rao, R. J., S. A. Hussain and R. K. Sharma (1989): The status and conservation of Ganges dolphin (*Platanista gangetica*) in the National Chambal Sanctuary. *Tiger paper*, 6–10.

Reeves, R. R. and R. L. Brownell, JR. (1989): Susu *Platanista gangetica* (Roxburgh, 1801) and *Platanista minor* (Owen, 1853). Pp 69–99. In: S. H. Ridgway and R. Harrison (eds.) *Handbook of Marine Mammals*. Vol. 4. Academic press, London.

Reeves, R. R., S. Leatherwood and R. S. L. Mohan (eds.) (1993): *A Future for Asian River Dolphins*: Report from the Seminar on the Conservation of River Dolphins of the Indian Subcontinent, 18–19 August 1992, New Delhi, India. Whale and Dolphin Conservation Society, Bath, U.K.

Reeves, R. R. and S. Leatherwood (eds.) (1995): Report of the First Meeting of the Asian River Dolphin Committee, Ocean Park, Hong Kong, 5-7 December 1994. Ocean Park Conservation Foundation. Honkong.

Reeves, R. R., B. D. Smith and T. Kasuya (eds.) (2000): *Biology and Conservation of Freshwater Cetaceans in Asia*. IUCN, Gland, Switzerland and Cambridge, UK, 152 pp.

Roxburgh, W. (1801): An account of a new species of *Delphinus*, an inhabitants of Ganges. *Asiatic Res*. (Calcutta), 7: 170–174.

Sharma, R. K. (1993): Annual gharial census in National Chambal Sanctuary (1992–93). *Journal of Tropical Forests*, 9(1): 90–93.

Shrestha, T. K. (1989): Biology, status and conservation of the Ganges river dolphin *Platanista gangetica* in Nepal. pp. 70-76. In: W.F. Perrin, R.L. Brownell Jr., Zhou kaiya and Liu Jiankang (eds.) *Biology and Conservation of the River Dolphins*. IUCN Species Survival Commission Occasional Paper No. 3. IUCN, Gland, Switzerland.

Shrestha, T. K. (1995): *The Ganges river dolphin*. Variety Printers Pvt. Ltd., Kathmandu, Nepal. 242 pp.

Singh, L. A. K. and R. K. Sharma (1985): Gangetic dolphin, *Platanista gangetica*, observations on habits and distribution pattern in National Chambal Sanctuary. *J. Bombay Nat. Hist. Soc*., 82: 648–653.

Sinha, R. K. (1997): Status and conservation of Ganges River Dolphin in Bhagirathi-Hooghly River Systems in India. *Internatl. J. Ecol. Environ. Sci*., 23: 343–355.

Sinha, R. K., B. D. Smith, G. Sharma, K. Prasad, B. C. Choudhury, K. Sapkota, R. K. Sharma and S. K. Behera (2000): Status and distribution of the Ganges susu. *Platanista Gangetica*, in the Ganges river system of India and Nepal. In: *Biology and Conservation of Freshwater Cetaceans in Asia*, Occasional Papers of the IUCN Species Survival Commission no. 23 (eds R. R. Reeves, B. D. Smith and T. Kasuya), pp. 54–61. IUCN, Gland, Switzerland.

Sinha, R. K. and G. Sharma (2003): Current status of the Ganges river dolphin, *Platanista gangetica* in the rivers Kosi and Son, Bihar, India. *J. Bombay nat. Hist. Soc*., 100(1): 27–37.

Sinha, R. K., N. K. Das, N. K. Singh, G. Sharma and S. N. Ahsan (1993): Gut content of the Gangetic dolphin, *Platanista gangetica*. *Invest. Cetacea*, 24: 317–321.

Smith, B. D. (1993): 1990 status and conservation of the Ganges River Dolphin (*Platanista gangetica*) in the Karnali River, Nepal. *Biological Conservation*, 66: 159–169.

Smith, B D., A. Benazir, A. Muhammad Edrise and Gill Braulik (2001): Status of the Ganges river dolphin or shushuk *Platanista gangetica* in Kaptai Lake and the southern rivers of Bangladesh. *Oryx*, 35 (1): 61–72.

Smith, B. D., A. K. M., Haque, M. S. Hussain and A. Khan (1998): River dolphins in Bangladesh: conservation and the effects of water development. *Environmental Management*, 22: 323–335.

Smith, B. D., R. K. Sinha, U. Regmi and K. Sapkota (1994): Status of Ganges river dolphin (*Platanista gangetica*) in the Karnali, Mahakali, Narayani and Saptakosi rivers of Nepal and India in 1993. *Marine Mammal Science*, 10: 368–375.

Smith, B. D. and R. R. Reeves (2000): Survey methods for population assessment of Asian river dolphins. In *Biology and Conservation of Freshwater Cetaceans in Asia*, Occasional Papers of the IUCN Species Survival Commission no. 23 80. pp. 97–115. IUCN, Gland, Switzerland.

Sterndale, R. A. (1884): *Natural history of the Mammalia of India and Ceylon*. Thacker, Spinkl and Co., Calcutta, Thacker and Co., Bombay, W. Thacker and Co., London, 540 pp.

Sinha, R.K., Verma, S.K. and Singh, Laliji (2010): Population status and conservation of the Gangetic dolphin (*Platanista Gangetica*) in the Indian Subcontinent. Chapter 22 In "Biology Evolution and conservation of River Dolphins with in South America and Asia. (Eds M. Ruiz- Garcia and Shostell) Nova Science Publishers., Inc. (New York, USA) ISBN 1608766330: 9781608766338. Page 419–443.

Singha, Th. P., Dutta, B.K. AND S.P. Biswas (2013): The Status of the Ganges River Dolphin *PLATANISTA GANGETICA GANGETICA* in the River Barak, ASSAM, INDIA *Journal of the Bombay Natural History Society*, 110(2), May–Aug 129–134.

Verma, S. K., R. K. Sinha and L. Singh (2004): Phylogenetic position of *Platanista gangetica*: insights from the mitochondrial cytochrome *b* and nuclear interphotoreceptor retinoid-binding protein gene sequences. *Molecular Phylogenetics and Evlution*, 33: 280–288.

Wakid, A (2005): Studies on certain Aspects of Ecology and Behaviour of River Dolphin (*Platanista gangetica*) in Dibru-Saikhowa National Park, Assam. Ph. D. Thesis, under Dibrugarh University, Assam. pp. 151 + plates.

Wakid, A (2006): Gangetic Dolphin Research and Conservation Programme, Aaranyak, 50, Samanwoy Path, Survey, Beltola, Guwahati 781 028, India. research article). *Current Science*, Vol. 97, No. 8.

Yeung, L.W.Y.,Yamashita,N., Taniyasu, S., Lam, K.S., D.V. and Kannan, K. (2009): A Survey of perfluorinated compounds in surface water and biota including dolphins from the Gangetic River and in other water bodies in India. Chemosphere. Vol. 76 (1): 55–62.

Chapter 6

Endangered Bird's Conservation

Ritu Agrawal and Sandeep P. Jadhav

College of Fisheries, GBPUAT,
Pantnagar – 263 145, Uttarakhand
E-mail: ritu_fish@rediffmail.com

ABSTRACT

Bird conservation is a field in the science of conservation biology related to threatened birds. Humans have had a profound effect on many bird species. Over one hundred species have gone extinct in historical times, although the most dramatic human-caused extinctions occurred in the Pacific Ocean as humans colonised the islands of Melanesia, Polynesia and Micronesia, during which an estimated 750-1800 species of bird went extinct.

According to World watch Institute, many bird populations are currently declining worldwide, with 1,200 species facing extinction in the next century. The biggest cited reason surrounds habitat loss. Other threats include overhunting, accidental mortality due to structural collisions, long-line fishing bycatch, pollution, competition and predation by pet cats, oil spills and pesticide use and climate change. Governments, along with numerous conservation charities, work to protect birds in various ways, including legislation, preserving and restoring bird habitat, and establishing captive populations for reintroductions.

Keywords: *Birds, Conservation, Endangered, Extinction.*

Threats to Birds

Habitat Loss

The most critical threat facing threatened birds is the destruction and fragmentation of habitat. The loss of forests, plains and other natural systems into agriculture, mines, and urban development, the draining of swamps and other wetlands, and logging reduce potential habitat for many species. In addition the remaining patches of habitat are often too small or fragmented by the construction of roads or other such barriers that cause populations in these fragmented *islands* to become vulnerable to localised extinction. In addition many forest species show limited abilities to disperse and occupy new forest fragments. The loss of tropical rainforest is the most pressing problem, as these forests hold the highest number of species yet, are being destroyed quickly. Habitat loss has been implicated in a number of extinctions, including the Ivory-billed Woodpecker (disputed because of "rediscovery"), Bachman's Warbler and the Dusky Seaside Sparrow.

Introduced Species

Historically the threat posed by introduced species has probably caused the most extinctions of birds, particularly on islands. Ninety per cent of historical extinctions have occurred on islands, and most prehistoric human caused extinctions were insular as well. Many island species evolved in the absence of predators and consequently lost many anti-predator behaviours. As humans travelled around the world they brought with them many foreign animals which disturbed these island species. Some of these were unfamiliar predators, like rats, feral cats, and pigs; others were competitors, such as other bird species, or herbivores that degraded breeding habitat. Disease can also play a role; introduced avian malaria is thought to be a primary cause of many extinctions in Hawaii. The Dodo is the most famous example of a species that was probably driven to extinction by introduced species (although human hunting also played a role), other species that were victims of introduced species were the Stephens Island Wren, Poo-uli and the Laysan Millerbird. Many species currently threatened with extinction are vulnerable to introduced species, such as the Kôkako, Black Robin, Mariana Crow, and the Hawaiian Duck.

Hunting and Exploitation

Humans have exploited birds for a very long time, and sometimes this exploitation has resulted in extinction. Overhunting occurred in some instances with native species unfamiliar with humans, such as the moa of New Zealand. In other cases it was an industrial level of hunting that led to extinction. The Passenger Pigeon was once the most numerous species of bird alive (possibly ever), overhunting reduced a species that once numbered in the billions to extinction. Hunting pressure can be for food, sport, feathers, or even come from scientists collecting museum specimens. Collection of Great Auks for museums pushed the already rare species to extinction.

The harvesting of parrots for the pet trade has led to many species becoming endangered. Between 1986 and 1988 two million parrots were legally imported into the US alone. Parrots are also illegally smuggled between countries, and rarer species can command high prices.

Hybridization

Hybridization may also endanger birds, damaging the gene stock. For example, the American Black Duck has been often reported hybridising with the Mallard, starting a slow decline.

Gamebird hybrids are particularly common and many breeders produce hybrids that may be accidentally or intentionally introduced into the wild.

Other Threats

Birds face a number of other threats. Pollution has led to serious declines in some species. The pesticide DDT was responsible for thinning egg shells in nesting birds, particularly seabirds and birds of prey that are high on the food chain. Seabirds are also vulnerable to oil spills, which destroy the plumage's waterproofing, causing the birds to drown or die of hypothermia.Light pollution can also have a damaging effect on some species, particularly nocturnal seabirds such as petrels.

Seabirds face another threat in the form of bycatch, where birds in the water become tangled in fishing nets or hooked on lines set out by long-line fisheries. As many as 100,000 albatrosses are hooked and drown each year on tuna lines set out by long-line fisheries.

Birds are also threatened by high rise buildings, communications towers, and wind farms; estimates vary from about 3.5 to 975 million birds a year in the North America alone. The largest source of human-related bird death is due to glass windows, which kill 100–900 million birds a year. The next largest sources of human-caused death are hunting (100+ million), house cats (100 million), cars and trucks (50 to 100 million), electric power lines (174 million), and pesticides (67 million). Birds are also killed in large quantities by flying into communication tower guidelines, usually after being attracted by tower lights. This phenomenon is called tower kill and is responsible for 5–50 million birds deaths a year. The presence of towers may seriously impact endangered species living in the vicinity.

Extinction of Birds in India

A number of birds are on the verge of extinction in the country. The International Union for Conservation of Nature (IUCN) red list of the birds indicates 15 species of birds from India as critically endangered. The information about the critical endangered birds is available on the IUCN Red list version 2013.

Bird species reported to be assigned under the Critically Endangered list of IUCN Red list are:

Sl.No.	*Common Name*	*Scientific Name*
1.	Baer's Pochard	*Aythyabaeri*
2.	Forest Owlet	*Heteroglauxblewitti*
3.	Great Indian Bustard	*Ardeotisnigriceps*
4.	Bengal Florican	*Houbaropsisbengalensis*
5.	Siberian Crane	*Grusleucogeranus*
6.	Spoon-billed Sandpiper	*Eurynorhynchuspygmeus*
7.	Sociable Lapwing	*Vanellusgregarius*
8.	Jerdon's Courser	*Rhinoptilusbitorquatus*
9.	White backed Vulture	*Gyps bengalensis*
10.	Red-headed Vulture	*Sarcogypscalvus*
11.	White-bellied Heron	*Ardeainsignis*

Sl.No.	Common Name	Scientific Name
12.	Slender-billed Vulture	*Gyps tenuirostris*
13.	Indian Vulture	*Gyps indicus*
14.	Himalayan Quail	*Ophrysiasuperciliosa*
15.	Pink-headed Duck	*Rhodonessacaryophyllacea*

It says that the reason for the decline in the number of these birds are:

- ☆ Loss, modification, fragmentation and degradation of habitat
- ☆ Environmental contaminants
- ☆ Poaching
- ☆ Use changes mainly conversion of large areas to intensive crop cultivation
- ☆ Changes in cropping pattern due to various reasons including implementation of irrigation schemes
- ☆ Increased pesticide usage and livestock-grazing
- ☆ High levels of disturbance
- ☆ Developmental activities like mining and hydel projects. This also includes collision of the birds with vehicles, wind turbines and power lines For conversion of the remaining habitats in the wet lands, grass lands and forests and river line across the countries, the steps taken by the government.
- ☆ For protection and management of protected areas the government of India provides financial and technical assistance to state/union territory government. The financial and technical support is also provided for forests protection under centrally sponsored schemes
- ☆ India has signed into several major international conventions related to conversion and management of wildlife, which includes the endangered species of birds
- ☆ The Central Government has enacted the Wildlife (Protection) Act, 1972 for protection of wildlife including birds. The Act, provides for creation of Protected Areas for protection of wildlife and also provides for punishment for hunting of specified fauna including birds specified in the schedules I to IV thereof. Important habitats of birds have been notified as Protected Areas under the Act
- ☆ Wetland (Conservation and Management) Rules 2010 have been framed for protection of wetlands, in the States, which are habitats of birds. The Centrally Sponsored Scheme of National Plan for Conservation of Aquatic Ecosystem also provides assistance to the States for management of wet lands including Ramsar sites in the country.
- ☆ Wildlife Crime Control Bureau has been established for control of illegal trade in wildlife, including endangered species of birds and their parts and products.

- ✰ Research and monitoring activities on birds are promoted by the Government through reputed research organizations. Wildlife Institute of India, Bombay Natural History society and Salim Ali Centre for Ornithology and Natural History are some of the research organizations undertaking research on conservation of birds.
- ✰ The Indian Government has banned the veterinary use of diclofenac drug that has caused rapid population decline of Gyps vulture across the Indian Subcontinent. Conservation Breeding Programmes to conserve these vulture species have been initiated at Pinjore (Haryana), Buxa (West Bengal) and Rani, Guwahati (Assam) by the Bombay Natural History Society.The important steps taken for protecting the endangered birds of India are given below:
  - (i) Rare and endangered birds are included in Schedule-I of the Wildlife (Protection) Act, 1972 thereby according them highest degree of protection.
  - (ii) Stringent punishments have been provided for in the Wildlife (Protection) Act, 1972 for violation of the provisions of the Act.
  - (iii) Important habitats of birds, including endangered and migratory birds have been notified as Protected Areas under the Wildlife (Protection) Act, 1972 for better conservation and the protection of birds and their habitats.
  - (iv) Financial and technical assistance is provided to the State/Union Territory Governments for protection and management of the Protected Areas.
  - (v) The State/Union Territory Governments have been requested to strengthen the field formations and intensify patrolling in and around the protected areas.
  - (vi Wildlife Crime Control Bureau has been established for control of illegal trade in wildlife, including endangered species of birds, and their parts and products.
  - (vii) India is signatory to all major international conventions relating to conservation and management of wildlife, including endangered species of birds. These are Convention on Biological Diversity, Convention on International Trade in Endangered Species of Wild Fauna and Flora, the Convention on International Trade in Endangered Species of Wild Fauna and Flora and Ramsar Convention on Wetlands of International Importance.
  - (viii) The Sub-committee constituted under National Board for Wildlife, has developed Guidelines for Threatened Species Recovery Plan and has also identified Great Indian Bustard, Jerdon's Courser and Nicobar Megapode as threatened bird species, for conservation and better management of their habitats on priority basis.

(ix) Research and monitoring activities on birds are promoted by the Government through reputed research organizations. Wildlife Institute of India, Bombay Natural History Society and Salim Ali Center for Ornithology and Natural History are some of the research organizations engaged in research for conservation of birds.

Conservation Techniques

Bird conservation practices are increasingly necessary to address the impacts of human activities that have accelerated extinctions and continue to threaten bird populations worldwide.

Habitat loss and fragmentation, overexploitation, chemical toxins and pollution, as well as introduced diseases, predators, and competitors are some of the threats that need to be addressed to reverse bird declines and maintain healthy populations.

The main objectives of bird conservation are to:

1. Identify population declines that signal underlying degradation of habitats and ecosystems
2. Identify the causes for these declines, and
3. Find biological solutions such as habitat preservation which can be used to repair the underlying causes of bird declines.

The trapping of birds has for long been a major problem in India. Munias, weavers, hill mynahs and parakeets are particularly sought after as cage birds. Many religious devotees take solace in releasing birds, in the mistaken belief that they are performing a 'good deed' by buying and setting them free. But in reality all they are doing is fuelling the trade. Most of these trapped birds are not naturally skilled to look after themselves, and when released find they are unable to fend for themselves and so die soon after. Meanwhile, the now richer bird-seller goes and captures some more to help assuage the conscience of the wannabe redeemer. Abrar Ahmed, who has contributed a substantial article on this problem writes "threatened birds are specifically trapped, traded and highly-valued". Species that fall in this category include the Swamp Francolin, Green Avadavat, Finn's Weaver, Sarus Crane, Black-necked Stork, Lesser Flamingo, and the Nicobar, Derby's, and Long-tailed Parakeets. All species of hornbills are highly prized and owls are in great demand for use in obscurantist tantric rituals. Numerous tribes (the Bahaleyiyas in North India, The Mirshikaris in the Northeast and the Narrikolvas, Hakkipukki and Harries in the peninsula) have for several centuries been traditional bird-trappers and unless we provide an alternate source of income for them, this scourge is likely to remain. Hunting, as a pastime by rich feudals and the then British rulers has certainly contributed to the decline of the big cats and other mammals in India. Bird hunting though was usually confined to ducks, pheasants, partridges, quails, Jungle fowl and snipe. There were certain rules that the 'sportsmen' followed: always shoot on the wing and invariably recover injured birds. This sport has now almost disappeared due to all hunting being banned, the difficulty in procuring gun licenses and the prohibitive expense of ammunition.

Scientists and conservation professionals have developed a number of techniques to protect bird species. These techniques have had varying levels of success

Captive Breeding

Captive breeding, or *ex situ* conservation, has been used in a number of instances to save species from extinction. The principal is to create a viable population of a species in either zoos or breeding facilities, for later reintroduction back into to the wild. As such a captive population can either serve as an insurance against the species going extinct in the wild or as a last-ditch effort in situations where conservation in the wild is impossible. Captive breeding has been used to save several species from extinction, the most famous example being the California Condor, a species that declined to less than thirty birds. In order to save the California Condor the decision was made to take every individual left in the wild into captivity. From these 22 individuals a breeding programme began that brought the numbers up to 273 by 2005. An even more impressive recovery was that of the Mauritius Kestrel, which by 1974 had dropped to only four individuals, yet by 2006 the population was 800.

Reintroduction and Translocations

Reintroductions of captive bred populations can occur to replenish wild populations of an endangered species, to create new populations or to restore a species after it has become extinct in the wild. Reintroductions helped bring the wild populations of Hawaiian Geese from 30 birds to over 500. The Mauritius Kestrel was successfully reintroduced into the wild after its captive breeding programme. Reintroductions can be very difficult and often fail if insufficient preparations are made, as species born in captivity may lack the skills and knowledge needed for life in the wild after living in captivity. Reintroductions can also fail if the causes of a birds decline have not been adequately addressed. Attempts to reintroduce the Bali Starling into the wild failed due to continued poaching of reintroduced birds.

The introduction of captives of unknown pedigree can pose a threat to native populations. Domestic fowl have threatened endemic species such as *Gallus g. bankiva* while pheasants such as the Ring-necked Pheasant and captive Cheer pheasants of uncertain origin have escaped into the wild or have been intentionally introduced. Green peafowl of similar mixed origins confiscated from local bird dealers have been released into areas with native wild birds.

Translocations involve moving populations of threatened species into areas of suitable habitat currently unused by the species. There are several reasons for doing this; the creation of secondary populations that act as an insurance against disaster, or in many cases threats faced by the original population in its current location. One famous translocation was of the Kakapo of New Zealand. These large flightless parrots were unable to cope with introduced predators in their remaining habitat on Stewart Island, so were moved to smaller offshore islands that had been cleared of predators. From there a recovery programme has managed to maintain and eventually increase their numbers.

Habitat Protection

As the loss and destruction of habitat is the most serious threat facing many bird species, conservation organisations and government agencies tasked with protecting birds work to protect areas of natural habitat. This can be achieved through purchasing land of conservation importance, setting aside land or gazetting it as a national park or other protected area, and passing legislation preventing landowners from undertaking damaging land use practices, or paying them not to undertake those activities. The goals of habitat protection for birds and other threatened animals and plants often conflicts with other stakeholders, such as landowners and businesses, who can face economically damaging restrictions on their activities. Plans to protect crucial habitat for the Spotted Owl of North America required the protection of large areas of old growth forest in the western United States; this was opposed by logging companies who claimed it would cause job losses and reduced profits.

Different Common Ways to Conserve Birds

It's easy to get involved in bird conservation, and like anything, some of your most helpful actions begin at home. Environment for the Americas surveyed biologists, educators, conservationists, birdwatchers, and bird enthusiasts to get their best ideas for getting involved in bird conservation. In celebration of International Migratory Bird Day's 20th Anniversary, we selected 20 issues and simple solutions. Imagine how many birds you can help finish their migratory journey, have a successful nesting season, raise young, survive the winter, if you just.

1. **Prevent Bird Collisions with Your Windows**: Collisions are one of the most frequent causes of bird deaths. Birds see nature reflected in the window or mistake houseplants inside the building for outdoor plants and fly into the glass. Putting up curtains or window decals helps make the window visible to birds.
2. **Protect Birds From Pets**: Unleashed dogs and outdoor cats can harm birds by disturbing, chasing, and even killing them. Keeping your cat indoors and your dog from straying saves millions of birds each year.
3. **Clean Your Bird Feeders:** Dirty feeders can spread disease. Disinfect and clean out old seed from feeders frequently and put fresh water in your bird bath every day.
4. **Don't Buy Illegally-Caged Birds:** Selling wild-caught birds as pets is illegal. Buy only captive-bred birds. Make certain that the breeder or pet store is reputable.
5. **Use Cloth Grocery Bags and Reusable Bottles:** Birds that mistakenly eat plastic trash can become ill or even die. Avoiding plastic bags and bottles reduces plastic pollution and conserves resources.
6. **Recycle**: Anything you recycle reduces litter and saves resources. Get creative! You can make a bag from used juice containers and bird feeders from old dishes and plastic bottles.

7. **Restore Natural Habitat in Your Community:** Birds need a place to live and many bird habitats are disappearing. City parks and open spaces are natural places for birds. Work with your community to recreate the habitat that once existed in your area.
8. **Keep Your Distance:** Birds need space for feeding, nesting, and other daily activities. Approaching too closely may cause them to become nervous and deplete much-needed energy reserves. During the nesting season, it may even result in loss of eggs or young to predators.
9. **Leave Fledglings Where You Find Them:** Fledglings may spend several days on the ground after they leave the nest before they are able to fly. You can help by keeping people and pets away, so their parents can continue to care for them. If you think a bird is truly an orphan, call a rehabilitator for instructions.
10. **Slow Down When Driving:** Cars kill millions of birds each year. Driving slowly gives you more time to respond if there is an animal in the road and gives the animal plenty of time to get out of the way.
11. **Buy Bird Friendly Products: We** can help preserve bird habitat by buying shade-grown coffee and chocolate. Shade coffee farms, which imitate native forests, have many more bird species than sun coffee farms.
12. **Plant Native**: Native plants provide food, nest sites, and cover for birds.
13. **Teach Others About Birds:** Talk to your friends about birds and start a club in your community to teach people about the challenges birds face. The more people know, the more they can do to help.
14. **Get Outdoors and Enjoy Nature: A**ppreciate the bird habitat near our home. Find a local park and go for a walk or just stroll around your neighborhood.
15. **Take a Friend Bird Watching:** Invite a buddy and see if you can spot more birds together.
16. **Support Conservation:** Join a bird club or other conservation organization to learn more and contribute to protecting birds. Volunteer with organizations that preserve habitat and help birds.
17. **Be a Citizen Scientist**: Many projects need helpers to gather data on birds and their habitat. Contact your local Nature Center, library or conservation organization to volunteer.
18. **Reduce Energy Use**: Riding your bike or walking reduces your carbon footprint and prevents pollution of bird habitats. Switching off the lights in your house not only shrinks your energy bill, but can also help prevent birds from colliding with your windows.
19. **Avoid Chemicals:** Birds may accidentally eat pesticide and herbicide pellets or prey that have been poisoned. This can kill a bird or have toxic effects on their own health and that of their growing embryos, including deformation or suppressed immune systems.

20. **Learn the Hunting Laws:** Federal and local laws protect sensitive areas and manage the harvest of birds to ensure healthy populations. Purchase a Federal Migratory Bird Hunting and Conservation Stamp. It contributes to habitat which benefits all birds.

References

1. Atkinson, C., Dusek, R., Woods, K., Iko, W. 2000. Pathogenicity of avian malaria in experimentally–infected Hawaii Amakihi. *Journal of Wildlife Diseases* 36(2): 197–204.
2. Blumstein, D., Daniel, J. 2005. The loss of anti–predator behaviour following isolation on islands. Proceedings of the Royal Society of London Series B–*Biological Sciences* 272: 1663–1668.
3. Brothers NP. 1991. Albatross mortality and associated bait loss in the Japanese longline fishery in the southern ocean. *Biological Conservation* 55: 255–268.
4. Dunnet, G., Crisp, D., Conan, G., Bourne, W. 1982.Oil Pollution and Seabird Populations [and Discussion]. Philosophical Transactions of the Royal Society of London. B 297(1087): 413–427.
5. Eckert, Allan W. 1965. The Silent Sky: The Incredible Extinction of the Passenger Pigeon. Lincoln NE: IUniverse.com. ISBN 0–595–08963–1.
6. Gill, F. 1995. Ornithology. W.H Freeman and Company, New York. ISBN 0–7167–2415–4.
7. Grier, W. 1982. Ban of DDT and subsequent recovery of Reproduction in bald eagles." *Science* 218(4578): 1232–1235.
8. Help Migratory Birds Reach Their Destinations. *Retrieved* 2006–07–21.
9. Holdaway, R., Jacomb,C. 2000. Rapid Extinction of the Moas (Aves: Dinornithiformes): Model, Test, and Implications. *Science* 287(5461): 2250–2254.
10. Jones, C.G., Heck, W., Lewis, R.E., Mungroo, Y., Slade, G., Cade, T. 1995. The restoration of the Mauritius kestrel Falco punctatus population. *Ibis* 137(Suppl.1): 173–180.
11. Klem, Jr., Daniel.1990.Collisions between birds and windows: mortality and prevention. *Journal of Field Ornithology.* 61 (l): 120–128.
12. Le Correa, M., Ollivier, A., Ribesc S., Jouventin, P. 2002. Light–induced mortality of petrels: a 4–year study from Réunion Island (Indian Ocean). *Biological Conservation* 105: 93–102 [1].
13. Moore, R., Robinson, W., Lovette, I., and Robinson, T. 2008. Experimental evidence for extreme dispersal limitation in tropical forest birds. *Ecology Letters,* 11(9): 960–968. doi: 10.1111/j.1461–0248.2008.01196.x
14. Philip J. K. McGowan, Peter J. Garson. 1995. Status Survey and Conservation Action Plan 1995–1999: Pheasants. Pheasant Specialist Group, IUCN. p. 7. ISBN 2–8317–0299–2.

15. Protect Backyard Birds and Wildlife: Keep Pet Cats Indoors. Retrieved 2006–07–21.
16. Putra, M. and Prins, H. 2000. Status and distribution of the endemic Bali starling *Leucopsar rothschildi*. *Oryx* 34(3): 188–197.
17. Simberloff, D. 1987. The Spotted Owl Fracas: Mixing Academic, Applied, and Political Ecology. *Ecology* 68(4): 766–772.
18. Steadman D. 2006. Extinction and Biogeography in Tropical Pacific Birds, University of Chicago Press. ISBN 978–0–226–77142–7.
19. Top 13 Killers. Fatal Light Awareness Program. *Retrieved* 2009–05–06.
20. Worldwatch Paper #165: Winged Messengers: The Decline of Birds. Worldwatch Institute. *Retrieved* 2006–07–21.

Chapter 7

Wildlife in India: Importance and Conservation Efforts

Akansha Khati

College of Fisheries, GBPUAT,
Pantnagar – 263 145, Uttarakhand

ABSTRACT

India's wildlife is rich and, diverse and awakes interest of people all over the world. Almost 4 per cent of India's land is under forests. India is the seventh largest country in the world and Asia's second largest nation with an area of 3,287,263 km^2, a national border of 15,200 km, and a coastline of 7516 km. For administrative purposes, India is divided into 28 states and union territories and is home to more than 1 billion people, which is approximately 16% of the world's population. Ecologically, India can be divided into three main regions, the Himalayan Mountain system, the peninsular India subregion (woodlands and desert); and the tropical rain forest region. The popular animals found in India are Apes, Asiatic Lions, Monkeys, Snakes, Crocodiles, Buffaloes, Elephants, Spotted Deer, Sambar Deer, Hog Deer, Barking Deer, Swamp Deer, One-Horned Rhinoceros, Sloth Bears, Blue Bull (Neelgai), Indian Gazelle (Chinkara), Goats and Tigers. Elephants provide maximum excitement while exploring India. India is also famous for its incredible tigers and has the largest number of bird species. There are more than 500 wildlife sanctuaries in India. Some of the major and popular wildlife sanctuaries are Corbett National Park, Gir Wildlife Sanctury, Kaziranga National Park in Sikkim, Periyar Wildlife Sanctuary, Neyyar Sanctuary, Blackbuck National Park, Kutch Desert Wildlife Sanctuary. The wildlife sanctuaries in India are home to around more than two thousand species of birds, three thousand five hundred species of mammals, nearly thirty thousand varieties of insects and more than fifteen thousand varieties of plants. There are as many as 80 national parks and over 441 wildlife sanctuaries in India, covering nearly 4.5 per cent of the total geographical area of the country.

Keywords: *Biodiversity, Conservation, Habitat, Natural Resources.*

India, lying within the Indo-Himalaya ecozone, is home to about 7.6 per cent of all mammalian, 12.6 per cent of avian, 6.2 per cent of reptilian, and 6.0 per cent of flowering plant species. 7.6 per cent mammals, 12.6 per cent aves, 6.2 per cent reptiles and 6 per cent flowering varieties of plant mark the biodiversity of India. Many ecoregions, such as the *shola* forests, also exhibit extremely high rates of endemism; overall, 33 per cent of Indian plant species are endemic. India's forest ranges from the tropical rainforest of the Andaman Islands, Western Ghats, and Northeast India to the coniferous forest of the Himalayas. Among these extremes lie the sal-dominated moist deciduous forest of eastern India; teak-dominated dry deciduous forest of central and southern India; and the babul-dominated thorn forest of the central Deccan and western Gangetic plain. Important Indian trees include the medicinal neem, widely used in rural Indian herbal remedies. The pipal fig tree, shown on the seals of Mohenjodaro, shaded the Gautama Buddha as he sought enlightenment. Neem and pipal fig tree are few of the important trees which have medicinal as well as asthetic value. Neem tree is considered as herbal home remedy for various diseases and pipal is linked with Gautam Buddha's enlightenment.

India is well known for its endangered and rare wildlife all over the world. Indian wildlife and vegetation is unique in its richness and diversity. India comparing to other countries has rare species in its jungles, wildlife national parks and sanctuaries. India's wildlife is both rich and varied. India has unique, rich and varied vegetation and wildlife. A survey says that 4 per cent of Indian land is covered with forests. India contains approximately 92 wildlife national parks and 480 wildlife sanctuaries. Here in this country, you may also find those species which are rare to find in any other part of the world. India is having 661 Protected Areas, which includes 100 National Parks, 514 Wildlife Sanctuaries, 43 Conservation Reserves and 4 Community Reserves.

In recent decades, human encroachment has posed a threat to India's wildlife; in response, the system of national parks and protected areas, National parks and protected area were first established in year 1935 and was substantially expanded. In year 1972, India enacted the Wildlife Protection Act and Project Tiger to safeguard crucial habitat; further federal protections were promulgated in the 1980s. The varied diverse and rich wildlife of India has had a profound impact on the region's popular culture. The common name for wilderness in India is Jungle, which was adopted into the English language. The word has been also made famous in The Jungle Book by Rudyard Kipling. India's wildlife has been the subject of numerous other tales and fables such as the Panchatantra.

Importance of Wildlife

Biodiversity

Biodiversity refers to the variety of life in any given area. Humans have identified 1.7 million species (groups of living organisms that interbreed) but estimates of total number of species on the planet go up to 100 million is seen in the number of species in an ecosystem or on the entire Earth. Biodiversity gets used as a measure of the health of biological systems, and to see if there is a danger that too many species become extinct. Tropical rainforests contain the most biodiversity, and invertebrates (animals without backbones) comprise half of the world's identified species.

Protections-via Legal Means

In the US, the Endangered Species Act designates threatened and endangered species and sets up programmes to protect their habitats. Globally, the Convention on International Trade in Endangered Species of Wild Fauna and Flora (CITES) unites the international community in protecting endangered species from human exploitation. These laws and others like them recognize the value of wildlife and represent important steps towards protecting biodiversity.

Human Impacts

The interconnectedness and interdependence of all the species in an ecosystem, the elimination of just one can have drastic and sudden effects on those remaining. There are five known mass extinctions in the earth's long history and many scientists say that humans are creating the sixth mass extinction today. This mass extinction is happening through the legal and illegal trade in at-risk species, expanding cities which destroy habitats, and human-driven climate change. These changes are happening so rapidly that many animals cannot adapt.

Wildlife Conservation

The need for conservation of wildlife in India is often questioned because of the apparently incorrect priority in the face of direct poverty of the people. However, Article 48 of the Constitution of India specifies that, "The state shall endeavor to protect and improve the environment and to safeguard the forests and wildlife of the country" and Article 51-A states that " it shall be the duty of every citizen of India to protect and improve the natural environment including forests, lakes, rivers, and wildlife and to have compassion for living creatures."

Large and charismatic mammals are important for wildlife tourism in India, and several national parks and wildlife sanctuaries cater to these needs. Project Tiger, started in 1972, is a major effort to conserve the tiger and its habitats. At the turn of the 20th century, one estimate of the tiger population in India placed the figure at 40,000, yet an Indian tiger census conducted in 2008 revealed the existence of only 1,411 tigers. 2010 Tiger census revealed that there are 1700 tigers left in India. The passing of the Forest Rights Act by the Indian government in 2008 has been the final nail in the coffin and has pushed the Indian tiger to the verge of extinction. Various pressures in the later part of the 20th century led to the progressive decline of wilderness resulting in the disturbance of viable tiger habitats. At the International Union for the Conservation of Nature and Natural Resources (IUCN) General Assembly meeting in Delhi in 1969, serious concern was voiced about the threat to several species of wildlife and the shrinkage of wilderness in India. In 1970, a national ban on tiger hunting was imposed, and in 1972 the Wildlife Protection Act came into force. The framework was then set up to formulate a project for tiger conservation with an ecological approach. However, there is not much optimism about this framework's ability to save the peacock, which is the national bird of India. George Schaller wrote about Tiger conservation.

Need for Importance of Conservation

Wildlife conservation has become an increasingly important practice due to the negative effects of human activity on wildlife. The science of extinction is called dirology. An endangered species is defined as a population of a living species that is in the danger of becoming extinct. The gradual emergence of the human beings as the most dominant species among all other species of animals and the attempt of the human beings to set them apart from other species is the main underlying cause of the contemporary environmental disaster. The main reason behind a threat to the wildlife and the ecosystem is the constantly growing deforestation, poaching and negligence towards animals and nature are habitat removal and alteration, overharvesting/overexploitation, pollution, introduction of exotic species, climate change, agriculture, fishing, forestry, urbanization, manufacturing and globalization are the major causes effecting biodiversity as a whole.

At the present estimate, 81 species of mammals, 38 species of birds, 18 species of amphibians and reptiles considered to be endangered in India. The tiger is the largest living member of the cat family, followed by the lion and the leopard. Habitat destruction and poaching brought about a sharp decline in their number and the national census of tigers in 1972 recorded that there were just 1827 of them in our country.

With the entire gloomy picture in regard of our wildlife, India is keen to do its best to protect its wildlife. Fortunately, we have ability and media, vocal environmental groups, NGOs and others who would not tolerate any more interference or intuition are working with full dedication to preserve and protect the vast diversity of animal wildlife.

Conservation Authorities

1. Important Nature Conservancy

The Nature Conservancy works with local communities, businesses, and individuals to protect over 100 million acres of land around the world. In doing so, The Nature Conservancy preserves entire wildlife communities and the rich species diversity that inhabits those lands. It's a holistic approach, one that is vital to the health of our planet.

Among the Nature Conservancy's more innovative conservation approaches is the debt-for-nature swaps. Such transactions ensure biodiversity conservation in exchange for debt owed by a developing country. Such debt-for-nature programmes have been successful in many countries including Panama, Peru, and Guatamala.

2. World Wildlife Fund

The World Wildlife Fund (WWF) is aimed at protecting biodiversity on a global scale. The WWF works with multilateral and bilateral agencies to promote sustainable development in the world's poorest countries. Its aims are threefold–to protect natural areas and wild populations, to minimize pollution, and to promote efficient, sustainable use of natural resources.

The WWF focuses their efforts at multiple levels, starting with wildlife, habitats and local communities and expanding up through governments and global networks. The WWF views the planet as a single, complex web of relationships between species, the environment, and human institutions such as government and global markets.

3. Natural Resources Defense Council

The Natural Resources Defense Council (NRDC) is an environmental action organization that consists of lawyers, scientists, and professionals. The NRDC uses the law, science, and their wide network of members and activists to protect wildlife and habitats around the globe.

NRDC focuses on curbing global warming, creating clean energy, preserving wildlands, restoring ocean habitats, stopping the spread of toxic chemicals, and working towards greener living in China.

4. The Sierra Club

The Sierra Club was founded by naturalist John Muir in 1892. The Sierra Club is a grassroots organization that works to protect ecological communities, encourage smart energy solutions, and to create an enduring legacy for America's wildernesses. Its current initiatives include coal alternatives, limiting greenhouse emissions, clean energy, green transport, and protecting wildlife communities.

The Sierra Club is involved in issues such as *viz.*, environmental justice, clean air, clean water, global population, toxic waste, and responsible trade. The Sierra Club supports vibrant local chapters that enable club members to become involved in conservation work in their area.

5. International Crane Foundation

The International Crane Foundation (ICF) was established by founders George Archibald and Ron Sauey in 1973 on a horse farm in Baraboo, Wisconsin. The ICF works around the world to protect cranes and the habitats on which they depend. Although they focus on cranes, their work is valuable on a wider scale, giving insights into endangered species management, wetland ecology, habitat restoration, and the critical need for international cooperation.

The ICF provides education about cranes on three levels – local, national, and international. In addition to educating people about cranes, the ICF also conducts captive breeding and reintroduction of cranes into the wild.

6. Friends of Haleakala National Park

The Friends of Haleakala National Park conservation organization support a wide range of conservation projects to protect Hawaii's unique Haleakala National Park. Their efforts include educational activities, cultural projects, research and service projects. The Friends of Haleakala National Park strives to preserve the ecosystems of Haleakala National Park, to protect the Native Hawaiian cultural, and to preserve the area's scenic character.

Haleakala National Park is located atop Maui's Haleakala volcano and is home to more threatened and endangered species than any other national park in the United States. Among the park's endangered species is the Hawaiian state bird, the Nene. The Friends of Haleakala National Park offers an adopt-a-nene programme that raises funds to protect the endangered nene goose from a range of threats including invasive predators such as mongooses, feral cats and rats.

7. Royal Society for the Protection of Birds

The Royal Society for the Protection of Birds (RSPB) began in 1889 as an organization that opposed the inhumane use of exotic feathers in the fashion industry, particularly the use of exotic plumes to adorn the women's hats that were so much in vogue at the time. The RSPB's rules were straightforward-to discourage the mindless destruction of birds, to promote the protection of birds, and to refrain from wearing feathers of any bird.

The RSPB protects and restores habitat for birds and other wildlife, conducts recovery projects, researches problems facing bird populations, works with landowners and farmers to protect birds, and manages 200 nature reserves. Each year, the RSPB conducts the Big Garden Birdwatch survey, which is a great way for people in the UK to participate in a nation-wide bird count.

8. Oceana

Oceana preserves the very core of our planet's health by protecting its the oceans. Healthy, rich and sustainable oceans and their resources are must hope for healthy landscapes. Our oceans sustain life on dry land through their role in maintaining the atmosphere we breathe, the water we drink, and the food we eat.

9. Conservation International

Conservation International employs scientists and policy experts to balance healthy ecosystems with sustainable human use. Conservation International aims to help stabilize global climate, protect fresh water, and ensure human well-being. To achieve their goals they work this organization works both with indigenous peoples and non-governmental organization. Conservation International's primary initiatives include climate, fresh water, food, health, culture, and biodiversity.

10. Wildlife Conservation Society

The Wildlife Conservation Society (WCS) is working for the protection of animals and wildlife. The WCS supports zoos and aquariums while promoting environmental education and conservation of wild populations and their habitats. They also offer educational resources and a wide variety of conservation programmes. Their efforts are focused on a select group of animals including bears, big cats, elephants, great apes, hoofed mammals, cetaceans, and carnivores. Their conservation projects stretch around the globe and are at work in regions including Africa, Asia, Latin America, the Caribbean, North America, and throughout the world's Oceans.

The Wildlife Conservation Society was established in 1895 as the New York Zoological Society. Its mission was, and is, to promote wildlife protection, foster the study of zoology, and create a top-notch zoo. Today not one but five Wildlife Conservation Zoos exist in the State of New York, the Bronx Zoo, Central Park Zoo, Queens Zoo, Prospect Park Zoo, and the New York Aquarium.

Conservation Methods

1. Habitat Protection

Perhaps the greatest threat that faces many species is the widespread destruction of habitat. Deforestation, farming, over-grazing and development all result in irreversible changes – soil compaction, erosion, desertification, and alteration of local climatic conditions. Such land use practices vastly alter or even eliminate wildlife habitat. In areas where rare species are present, habitat destruction can quickly force a species to extinction.

By protecting habitat, entire communities of animals can be protected together and when communities are kept intact, less conservation intervention is required to ensure species survival. Parks, reserves, and other protected lands are too often the only habitats left untouched by habitat destruction.

2. Control of Exotic Species

The spread of non-native species has greatly impacted native populations around the world. Invasive species compete with native species for resources and habitat. They can even prey on native species directly, forcing native species towards extinction.

3. Recycling and Energy Reduction

By recycling and (reusing as much as we can, we reduce our impact on the environment. Additionally, by reducing the energy we consume, we take a little of the burden off our natural resources.

4. Reduction of Pesticide and Herbicide usage

Herbicides and pesticides may keep yards looking nice but they are in fact hazardous pollutants that affect wildlife at many levels. Many herbicides and pesticides take a long time to degrade and build up in the soils or throughout the food chain. Some groups of animals such as amphibians are particularly vulnerable to these chemical pollutants and suffer greatly as a result of the high levels of herbicides and pesticides in their habitat.

In order to instigate wildlife conservation efforts in India, World Wildlife Fund has set up various groups. India hosts diverse kinds of wildlife species possibly due to the broad range of weather, soil, climate, and other similar factors. On the negative side, however, there are certain species that are threatened or vulnerable and are required protection. Perhaps the need of the time is to put more and more conservation efforts to protect wildlife in India.

Wildlife Conservation Efforts in India

Today, India contains 172 species (2.9% of the world's total number) of animals that are considered to be globally threatened by the IUCN. These include 53 species of mammals, 69 species of birds, 23 species of reptiles, and 3 species of amphibians, human encroachment is one of the biggest threats to India's wildlife. Indian Government built many National Parks in order to facilitate endangered species up to some extent. The first park was established in 1935; besides, the Project Tiger was also enacted in 1972. At the beginning of the 20th century, there were around 40,000 tigers in India as compared to the tiger consensus of 2008 which revealed only 1,411 tigers remaining. The 'Project Tiger' is considered to be the most successful conservation project all throughout India. Currently, there are around 39 Project Tiger wildlife reserves in India encompassing an area of more than 37,761 sq. km. Many rhino's are survived and conserved in the Kaziranga National Park.

- The Forest Conservation Act 1980 was enacted to lessen the large-scale habitat destruction caused by deforestation.
- India also indorsed the Convention on International Trade in Endangered Species of Wild Fauna and Flora (CITIES) in 1976. As a result, ban was imposed on the commercial trading of Asian elephants which involved the export of ivory.
- There has been a significant change in the Forest Department's planning policy, which not only brought an end to the alteration of first-growth forests into eucalyptus plantations, but it also swung the importance from a commercially-centered forest management to the conservation-centered forest management.
- The elephants have enjoyed an absolute protection in Uttar Pradesh, since the promulgation of a Wild Elephant Protection Act 1879. The authorities are also putting their efforts to instigate Rajaji National Park, Motichur and Chilla reserves.
- The Forest Department also compensated the families whose persons were killed by the wild animals.
- The capturing of elephants has been superseded.
- A new conservation area was also established which was named as Nilgiri Biosphere Reserve covering 5,520 km^2. This area protects elephants ranging from North of Palghat Gap in Karnataka, Tamilnadu, and Kerala.
- The Karnataka state government has decided to exclude tourist development on elephant migration routes all throughout the Kabbini River.

Wildlife Conservation: Some Case Studies

Shahtoosh Case

The Shahtoosh wool is derived from the soft undercoat of the Tibetan Antelope (also known as Chiru), which has to be killed before its fleece is removed. Three to four Chiru have to be killed to weave only one shawl. Each shawl can cost several thousand dollars in the international market.

In 1977, the Government of India declared the Chiru as protected under Schedule I of the Wildlife (Protection) Act of India, 1972. Killing of Chiru is also in contravention to the Convention on International Trade in Endangered Species (CITES), of which India is also a signatory as mentioned earlier.

A PIL or Public Interest Litigation was filed in the J and K High Court seeking implementation of the provisions of their Wildlife (Protection) Act as well as CITES which prohibits the import of Shahtoosh into India. On May 1, 2000, the Honorable High Court issued a judgment forcing the government to enact and enforce its wildlife law. Finally in 2002 the manufacture of Shahtoosh shawls has finally been banned in the state of Jammu and Kashmir.

Project Tiger

Project Tiger, an initiative by the Government, launched on April 1, 1973, has become one of the most successful conservation ventures in modern history. The project aims at tiger conservation in specially constituted 'tiger reserves' which are representative of various bio-geographical regions falling within India. It strives to maintain a viable tiger population in their natural environment. Today, there are 27 Project Tiger wildlife reserves in India covering an area of 37,761 km.

Conclusion

Great Indian Bustard

The great Indian bustard (*Ardeotis nigriceps*) is a large, handsome bird of the shortgrass plains of the Indian subcontinent. The Bombay Natural History Society is a pioneer in promoting the conservation of the great Indian bustard, and undertakes intensive campaigns to educate and encourage the Government of India to take appropriate measures to reverse the declining trend of the species.

Freshwater Gangetic Gharial

The gangetic gharial (*Gavialis gangeticus*) is a thoroughly aquatic crocodilian and a resident of deep, fast flowing rivers. It is primarily a fish-eating species and uses sandbanks for nesting. The gangetic gharial used to have wide range over all of Indochina, but today it is the most endangered of all the crocodilians. In India, there are nine protected areas, with a total area of nearly 3000 km2 that are designated for gharial management.

Wildlife habitat and species around the world are facing severe crisis. It is estimated that global warming may cause the extinction of 15–37 per cent of species by 2050. This is another aspect which needs attention because we could lose about 1.25 million species. Unlike other environmental losses, this one cannot be reversed because nature does not give second chances to biodiversity. India has launched an extensive protected area network of research institutions in which legislation, socio-economic factors, and wildlife research are playing a great role. The Central Zoo Authority plays a key role with zoos in programming research activities related to the conservation and propagation of wild animals. Wild life protection is not possible by laws only. Active cooperation and participation is must from all of us. As it is high time to understand the solemnity of situation and act responsibly. It can be achieved by strict enforcement of laws and mass awareness. The future depends on interaction between captive and wild animals, preservation of biodiversity, and genetic conservation of species.

References

Milton L (2013). Karnataka witnesses highest elephant deaths. *The Times of India* February 6, 2013..

Sutirtha Dutta, Asad R Rahmani and Yadvendradev V Jhala (2013). Running out time? The great Indian Bustard *Ardeotis nigriceps* – Status, viability and conservation strategies. *European Journal of Wildlife Research* 57: 615-625.

Ullas Karanth K (1995). Estimating tiger Panthera tigris populations from camera-trap data using capture-recapture models. *Biology conservation* 71: 333-338.

Chapter 8

Conservation and Rehabilitation of Critically Endangered Gangatic Ghariyal (*Gavialis ganeticus*)

Amita Saxena[1], Anita[1] and Maneesh Kumar Dubey[2]

[1]Department of Fisheries Resource Management, College of Fisheries, GBPUAT, Pantnagar – 263 145, Uttarakhand

[2]Virology Laboratory, Directorate of Cold Water Fisheries, Bhimtal, Nainital

---

ABSTRACT

The Indohimalaya ecozone of India is home for about 7.6 per cent of all mammals, 6.0 per cent of flowering plants, 12.6 per cent of Avian population and 6.2 per cent of reptile population of the world. Gharials (Gavialis gangeticus) are one of the crocodile species native to India. Gharials are commonly known as gavial, Indian gharial etc. and scientifically named as Gavialis gengeticus. The gharial are also known as fish eating crocodiles or long snout crocodiles. They are native to the great Ganga river system of Indian Subcontinent. They live in the deep and flowing river system. They are listed as critically endangered animals by IUCN because of drastic decline in their population and human activities are responsible for such declination. Few years ago gharials inhabited to the Indian Subcontinent from Irrawaddy river in the east to the Indus river to the west but now their distribution is limited to only 2 per cent of the former range which is sign of huge reduction in population of this aquatic animal. Many initiatives have been taken by Wildlife reaesrchers and government to conserve and restore the breeding population of gavialis to the rivers. Pollution and disbalanced aquatic environment also negatively affecting the existence of this animal. The present article is focusing on habit habitat, current status, threats, conservation and rehabilitation of gharial (Gavialis gangeticus) in India.

Keywords: *Gharial, IUCN, Critically endangered, Rehablitation.*

Introduction

India is endowed with unique and rich biodiversity of wild animals or plants as it is the home for 60–70 per cent of the world biodiversity. India is one of the 17 mega diverse countries of the world. The country is also home for various animals like Bengal tiger, lions, deer, foxes, certain birds, monkeys, snakes, camels, bears, crocodiles, wolves, pythons, wild dogs, elephants, Indian rhinoceros and Indian whit- rumped vulture. Such unique and rich biodiversity of these animals is conserved in 400 and above wildlife sanctuaries, 18 Bioreserves, 26 wetlands are registered under Ramsar Convention and 89 National parks in the country. National parks and protected areas were first established in 1935 when Human encroachment posed threat to Indian wildlife biodiversity and India enacted wildlife protection act in 1972 for conservation and re habilitation of biodiversity. India has three most diverse biodiversity hotspots Western Ghats and Indo-Burma which are home for various threatened and endangered species of animals and plants.

The taxonomical classification of Gharial puts this animal in to Kingdom Animalia, Phylum Chordata, Class Reptilia, Order Crocodylia and family Gavialidae. Scientifically named *Gavialis gangeticus* is one of the three native crocodiles of Indian subcontinent Mugger crocodile and salt water crocodile. Gharial is also one of the longest living crocodile ranges upto the size of 6.25m and they also have 110 sharp interdigitated teeths in its long earthen pot ' ghara' like thin snout. Gharials inhabits in the flowing rivers which have large sand banks. They generally use these sand banks for basking activity and nest building. They usually breeds during the cold season and the once hatch out before the onset of monsoon season.

These animals are characterized by elongated and thin jaw (adapted for fish diet). Male reach up to the size of 6 m with an average weight of around 160 kg. The total estimated population of gharial is fewer than 235 individuals and their population is drastically decreasing since last 70 years so the animal is listed as *Critically Endangered* by IUCN. There are various reasons for such sharp decline of the population of these animals like destruction of riverine habitat, depletion of fish resources and use of destructive fishing nets. Heavy fishing activities, human activities, distruction of sand banks (basking), boating activities and pollution obstructs the spotting of gharials and they possess the main threat to this species. Gharial is true dweller and eats only fish. Such a narrow food and space choice is also responsible for drastic decline of these aquatic animals from Ganga river system.

Specific Identifying Characters

- ☆ Gharials are dark or light olive in colour with dark cross-bands and speckling on the head, body and tail.
- ☆ Dorsal surface of the body become dark, almost grey-black,as the age increases above 20 years while Ventrals are yellowish-white.
- ☆ The neck is elongated and thick and fingers are extremely short and thickly emarginated with a web.
- ☆ Males develop a hollow bulbous nasal protuderance which resembles like earthen pot 'ghara' during sexual maturity. It also acts as sound resonator and also indicates sexual behavior of this animal.

- ✰ The average size of mature gharials is 3.5 to 4.5 m while the largest recorded length is around 6.25 m (20.5 ft).
- ✰ The young gharials reach to a length of 1 m in 18 month and the average weight ranges to 159 to 250 kg.
- ✰ First maturity commences in the male gharials when they attain an average length of 3 to 5 m (9.8 to 16.4 ft), while females attains a body length of up to 2.7 to 3.75 m (8.9 to 12.3 ft).
- ✰ The elongated, narrow snout is well equipped with 110 sharp interdigitated teeth, and becomes proportionally shorter and thicker when the age of the animal grows.
- ✰ The laterally flattened tail and webbed rear feet assist in feeding and dwelling in deep flowing water habitat. Such unique structure of tail helps in both activities, *i.e.*, movement and as a base from which to strike at prey.

Habit and Habitat

Gavialis gangeticus (gharials) prefers to live in flowing riverine system with deep pool. They need sandy bed for nest building and basking. The young gharials feed upon insects, tadpols, small fishes and frogs while the adult gharials feed on fishes and small crustaceans. They do not eat human because their jaw is very thin and delicate to grab a human. They do not chew their prey, directly swallow it. Male grows faster than female and the adult once are unable to move semi up right procedure (Bustard and Singh. 1977).

Reproductive Biology

Males commence maturity at the age of 13 years old, when the 'ghara' begins to grow on the snout (Bustard and Maharana, 1982). They attract females for mating by making hissing and buzzing noises as they patrol their territories. Females show mating desire by raising their snouts upwards. During courtship both of them rub and mount the head and snout. The mating season of gharials is during December to January and they dig nests during March to April away from the waterline by using their hind feet. In which female lays 20–95 eggs of 160 g approximately largest eggs of any crocodilian species (Hussain, 1999). After 71–93 days of incubation the newly hatched young gavials emerge out from the eggs before the onset of monsoon season. The sex of the newly-hatched gharials is determined by the temperature. Unlike other animals these animals also shows parental care. The Female of the population guards the hatchlings but do not support them to reach the water.

Factors Causing Threat to Gharials

- ✰ The first and most important factor responsible for drastic decline in population of gharials is heavy hunting and smuggling teeths and skin of gharials.
- ✰ Distruction of large sand beds along the river system and high level of industrial pollution in the Ganga river system. As these sand bank serve as nest building grounds of gharials.

- Narrow choice of food (eats only fish) may be responsible for decline in the population as the fish resources are declining itself.
- Heavy hauman activities and disturbances also negatively affect the population of gharials.
- Predation of eggs of gharials by birds and other predatory animals is also responsible for reduction in the population.
- Use of destructive fishing gears that negatively affects the population of gavials.
- Construction of dams, barrages, artificial embankments, irrigation channels, Sand mining leads to siltation of rivers which seriously hampered the life cycle of gharials in the river system.
- Killing of gharials for medicine purpose by the local peoples.
- Consumption of eggs by human.

Current Status of *Gavialis gangeticus*

Gharials plays important role in aquatic ecosystem as they are tertiary consumer in the aquatic environment any hence they are very important for maintaining balance in aquatic food chain. The population of gharials in wild in 1976 was 5000 to 10,000 but the present status is less than 200 individuals showing a drastic decline of 96 per cent. Current estimates from population surveys carried out in 2007 indicated that only 200–300 breeding adults of *Gavialis gangeticus* is left in the world. Chambal river system posses the largest population of gharials around 306 adult gharials (Bhadoria *et al.*, 2009).

Recently the breeding populations of gharials are limited to only few tributaries of Ganga river system in India and the Narayani river system in Nepal. Recently small population of gharial is reported from the hoogly river system of West Bangal. Now in India only small populations of *Gavialis gangeticus* is present in the rivers of National Chambal Santuary, Katamiaghat wildlife sanctuary Son river sanctuary some part of Mahanadi (Satkosia Gorge Sanctuary) in Odisha but they apparentaly do not breed at these places. Small population of gharials is also recovering from the tributaries of Ganga river system like Narayani Rapti river system and Karnali Babai river system. Recently the wild population of gharails showed drastic decline of about 58 per cent during 1997–2006.

Conservation and Rehabilitation

Wildlife department of India is actively working in the field of conservation and rehabilitation of *Gavialis gangeticus* in Ganga river system. The wildlife Protection Act formulated in by the government of India in 1972 for protection of wildlife biodiversity. For conservation of these critically endangered animals government puts ban on capture and hunting of gharials from the Ganga river system. The policies are also formulated for protection purpose. In 1977, four nests were recorded in the Girwa River of Katarniaghat Wildlife Sanctuary, where 909 gharials were released until 2006. Twenty nests were recorded in 2006, so 16 nesting females resulted from 30 years of reintroductions, which is equivalent to 2 per cent of the total pre-2006 releases. Under rehabilitation programme in 1978, twelve nests were recorded in the National Chambal Sanctuary where 3,776 gharials were released until 2006. During the year 2006, nesting had increased to 68 nests. Although gharials were re habilitated in the sanctuaries but the recruited mature, reproducing females constituted only about 2 per cent of the total number released and the main obstruct of growth of young once is that the newly-hatched young are especially prone to being flushed downstream out of the protected areas during the annual monsoonal flooding which resulted in death of the young once.

The concerted efforts of government of India for restoration of gharial population are attempted in many states like Girwa in Uttarpradesh, Chambal in Uttarapradesh and Madhya Pradesh and Mahanadi in Odisha. Government of India is now stepping ahead in protection and rehabilitation of these critically endangered animals. The captive breeding centres for breeding and rearing of gharials is being established by the government. Under this prorgamme gavials are being bred in National Chambal Sanctuary in Uttar Pradesh, India. Nepal is also working in this field the 'Chitwan National Park' of Nepal grows gharials for rehabilitation purpose. Nepal also established gharial breeding centre. In India gharials are kept in Jawaharlal Nehru Biological Park, Madras Crocodile Bank, Indra Gandhi Zoological parks, Bannerghatta Zoo, Junagarh Zoo, Bilogical park of Itanagar and Kukrail Forest of Lucknow india.

Limitation of Breeding Success of Gharials

- The most important drawback is lack of breeding population in wild.
- The survey of species is only limited to protected areas only not in the un explored waters. So we needs to study scope of their restoration and rehabilitation in unexplored or under explored areas.

Conclusion

Gharial *Gavialis gangeticus* population is declining drastically due to the destruction of their natural habitat and reduction of fishery resources. The IUCN listed these Indian gavials as critically endangered animals. So the hunting, killing, trafficking of gharial's teeths, skin and other body parts should be banned. Artificial breeding grounds and basking places should be constructed for this species otherwise this gavialis species of gavialis which is native to Indian subcontinent will extinct from wild.

Rererences

Bhadoria, S. C., Luikham, N., and Sharma, R.K. 2009. Survey Report of Aquatic Animals in National Chambal Sanctuary, Morena, MP. Report to Gwalior Circle Gwalior.

Bustard, H. R., and L. A. K. Singh 1977. Studies on Indian Gharial *Gavialis gangeticus* (Gmelin): I. change in terrestrial locomotary pattern with age. *Journal of Bombay Natural History Society* 74 (3): 534–535

Bustard H. R. and S. Maharana 1982. Growth rate in subadult gharial *Gavialis gangeticus*, Gmelin, Reptilia, Crocodilia. *Journal of Bombay Natural History Society*. 80: 224–226.

Hussain, S. A. 1999. Reproductive success, hatchling survival and rate of increase ofgharial *Gavialis gangeticus* in National Chambal Sanctuary, India. *Biological Conservation* 87: 261–268.

IUCN/Species Survival Commission. 2008. Strategic Planning for Species Conservation: An Overview. Version 1.0. Gland, Switzerland: IUCN. 22 pp.

Stevenson, Colin and Whitaker, Romulus 2010. Gharial *Gavialis gangeticus*. Status Survey and Conservation Action Plan. Third Edition, ed. by S.C. Manolis and C. Stevenson. Crocodile Specialist Group: Darwin pp. 139–143.

Chapter 9

Present Status and Conservation of the Indian Rhinoceros (*Rhinoceros unicornis*)

Pratibha, Seema Jain and Nisha Rana

Department of Zoology,
R.G.P.G. College, Meerut, U.P.

ABSTRACT

The Indian rhinoceros (Rhinoceros unicornis), also called the "greater one-horned rhinoceros" or "Asian one-horned rhinoceros" belongs to the family Rhinocerotidae. It is a small species of rhinoceros. The Indian rhinoceros gets its common name from the fact that it has only one horn rather than two. Rhinoceros is from Greek words rhino for "nose" and ceros meaning "horn". unicornis is from Latin words uni meaning "one" and cornis meaning "horn"). The Indian rhinoceros has one horn which it uses for defence, intimidation, digging up roots and breaking branches during feeding. This animal has unique 'armour-plating' appearance, and they have a prehensile upper lip, much like that of the black rhino, which assists in foraging. Rhinoceros unicornis is the fifth largest and also the largest of the Asian rhino species. This species is now listed as a vulnerable species.

***Keywords**: Habitat, Rhinoceros, Wallowing.*

Introduction

The Indian rhinoceros (rhino) is primarily found in north-eastern India's Assam and in protected areas in the Terai of Nepal, where populations are confined to the riverine grasslands in the foothills of the Himalayas. The preferred habitat of an Indian Rhinoceros is alluvial flood plains and areas containing tall grasslands along the foothills of the Himalayas. Rhino populations were severely depleted as they were hunted for sport and killing as agricultural pests. This pushed the species very closely to extinction in the early 20th century and by 1975 there were only 600 individuals surviving in the world. In India, rhinos are found in Kaziranga, Orang, Pobitara, Jaldapara, Dudhwa. Thanks to rigorous conservation efforts, their numbers have increased dramatically since 1975. By 2012, conservation efforts saw the population grow to over 3,000 in the Terai Arc Landscape of India and Nepal, and the grasslands of Assam and north Bengal in north-east India.

Systematic Position

Phylum:	Chordata
Sub phylum:	Vertebrata
Super class:	Tetrapoda
Class:	Mammalia
Subclass:	Theria
Infra class:	Eutheria
Order:	Perissodactyla
Family:	Rhinocerotidae
Genus:	*Rhinoceros*
Species:	*unicornis*

External Morphology and Physical Features

Size and Weight

Male Indian rhinos weight approximately 2,200 kg (nearly 4,840 pounds) and range in height from 170–186 cm (67–73 inches) and are 368–380 cm (145–150 inches) long. Females are smaller, weighing only 1,600 kg (3527 pounds) and standing 148–173 cm (58–68 inches) tall. Female Indian rhinos are 310–340 cm (122–134 inches) long. The average height is about 170c m (5ft. 8 in.) with a girth of 335 cm (11 ft). While a fully grown male rhino weighs around 2000–2500 kg, a female weighs around 1600 kg.

Skin Colour

They have an ashy grey, hairless, knobby, leathery, hard and thick skin which develops thick folds, resembling armour plating. Several prominent folds of skin protect the neck. The skin has a maximum thickness of four cm; the subcutaneous fat is 2.5 cm thick and well supplied with blood, which helps in thermoregulation, so that the animal is able to regulate its own body temperature in varying weather conditions. Between the folds, around the stomach, the inner legs and the facial area, the skin is rather soft and thin.

Hairs

Hairs found only at the tip of the tail, around the ears and eyelids.

The Horn

A single horn sits on top of their snout, and their upper lip is semi-prehensile. The horn of the Indian rhinoceros is typically 20–61cm long, and weight up to

65 kg. The horn can grow upto 45 cm (85 inches). It has the same horn structure as the hooves of horse and re-grows if broken off. It is made from a substance called keratin and is, therefore, very strong.

Other Sense Organs

The Indian rhinoceros has relatively poor eyesight, relying more on hearing and smell to detect what is going on around them. The ears of the Indian rhinoceros possess a relatively wide rotational range to detect sounds and an excellent sense of smell to readily alert them to the presence of predators.

Teeth

Although their horn may not be as long as other well-known species of rhino, Greater one-horned rhinos have very long lower incisor teeth that can be used in fighting and to inflict deep wounds. In males they can grow up to 8 cm long.

Food and Feeding

Greater one-horned rhinos is herbivorus (they are primarily grazers) animal and eat on an average 1 per cent of their body weight daily, and are known to swim for their food. They sustain themselves on a purely plant based diet and feed on densely vegetated subtropical forest for fruits, flowers, buds, leaves and branches of trees and shrubs, grass, submerged and floating aquatic plants and agricultural crops, berries and roots which they dig up from ground using their horn.

Social Behaviour and Territorial Behaviour

Greater one-horned rhinos are usually solitary creatures, except when sub-adult or adult males gather at wallows or to graze and females with small calves. Males have loosely defined territories where they live alone, and which they defend aggressively, but may overlap with other territories. The territories change according to food availability in relation to the current season. The female moves in and out of these territories as they like. If food is abundant within an area, it is not unusual to see several animals all grazing close together.

Forest Paths and Scent-marking

They tend to use the same path, which are marked by a scent gland on the bottom of their feet. 'Middens' (rhino dung heaps) serve as communication points and mark territorial boundaries. Several animals often defecate at the same spot. Such a dung heap can become up to five metres wide and one metre high. After defecating, greater one-horned rhinos scratch their hind feet in the dung and continue to walk, they "transport" their own smell around the paths, thus establishing a scent-marked trail that is claimed by the rhino in question.

Wallowing

Mud wallows can be places where several individuals meet, as a kind of social gathering. Afterwards, they separate again. By covering their skin in mud, this aids thermoregulation by preventing overheating, and also suffocates any ticks or parasites that are embedded on the surface of the skin.

Sounds

Rhinoceros can communicate with 12 different, sounds, of which frequently used are snorts, honks and roars.

Running Speed

It can run 40-42 km/hr as fast as and are very agile, despite their bulky shape and size while it can run at speeds of up to 55 km/hr for short periods of time. It is also an excellent swimmer.

Swimming

They are very good swimmers and can dive and feed under water, seemingly enjoying the cool, wet elements of the surrounding lakes and riverines of their habitats.

Longevity

Greater one-horned rhinos lives on an average upto 30–45 years in the wild; while the longevity record for those in captivity is 40–50 years.

Breeding

Sexual Maturity

Breeding takes place at all times of the year. The Indian rhinoceros is solitary animal and only comes together with other Indian rhinos to mate. Females starts breeding when 4–5 years old, and males are usually sexually mature when 7–10 years old. A female will leave a gap of around 3–4 years between calves.

Gestation Period

This is between 15-16 months.

Birth

The female Indian rhinoceros gives birth to a single calf after a gestation period that is over a year long. Just as female is ready to give birth, the female will find a solitary, quiet space to calve. The calf remains with its mother for the first year and a half of its life, before being rejected. At birth, a greater one-horned rhino calf is around 105 cm long and weighs as much as 58–70 kg.

Parental Care/Maternal Calves

A calf drinks on average 20–30 litres of milk per day and grows by 1–2 kg daily. They start nibbling/feeding on roughage at the age of 3–5 months and continue to suck up to the age of 20 months in some cases.

Habit and Habitat

Greater one-horned rhinos are semi-aquatic, solitary and are often seen to take up residence in swamps, forests and riversides, tropical and sub tropical grassland, savannas, deserts and shrub land.

Distribution

This animal is found in India and Nepal and particularly in the foothills of the Himalayas. In the past, greater one-horned rhinos roamed freely in the alluvial floodplains, tall grassland and forests alongside the Brahmaputra, Ganges and Indus River valley. Today the species is restricted to small habitats in Indo-Nepal terai, North Bengal, and Assam.

Population Status

Today, the Indian rhinoceros is a vulnerable animal and has been pushed into only a small fraction of its historical territory by human hunters and deforestation. There are around 3,000 Indian rhinoceros individuals left in the wild, two thirds of which are believed to be in the Assam region of India(4). In 2014, 2,544 were recorded in Assam (Kaziranga National Park, Chitwan National Park) alone, an increase by 27 perc ent since 2006, although in early 1900s, Assam had about 200 rhinos only.

Population status in India: Kaziranga National Park [March 2012]–2290 rhinos, Pabitora Wildlife Sanctuary [March 2012] –93, Orang National Park [March 2012] –100, Manas Tiger Reserve [March 2012] –22 (translocated from Pabitora and Kaziranga since 2006). Currently 2,619 rhinoceros were recorded in wild whereas 158 Indian Rhino were recorded in captivity (At the end of 2006).

Reasons for Depletion and Threat

Today, the need for land by the growing human population is a threat to the species. Many of the protected areas where rhinos have now reached the limit of the number of individuals they can support. This animal is now facing several threats due to illegal hunting, poaching, loss of habitat, etc. Some of these are as follows:

Hunting and Poaching

Humans are the biggest threat to the Indian rhinoceros as they have been hunted to the brink of extinction for their horns as they are used in Asia as medicine for fever and pain. Historically, the Indian rhinoceros had a vast range across northern India but today that range has been drastically reduced due to excessive hunting. In the early 19th century, the greater one-horned rhino was almost hunted to extinction, rhinos were hunted for sport by both Europeans and Asians and also killed as agricultural pests in tea plantations. The remaining animals were only found in protected reserves, where under the monitoring of certain organisations, populations are currently being brought back from the edge of extinction.

Illegal Wildlife Trade

Despite protections and bans on international trade of rhino horn, extensive illegal trade persists throughout Asia. Although there is no scientific proof of its medicinal value, the horn is used in traditional Asian medicines, primarily for the treatment of a variety of ailments including epilepsy, fevers and strokes. Asian rhino horn is believed to be more effective than African horn.

Habitat Destruction

The enormous reduction in the range of rhinos was mainly caused by the disappearance of alluvial plain grasslands and deforestation. As Greater one-horned rhinos live in areas with very fertile soil, humans use the same land for farming purposes. Today, the need for land by growing human population is a threat to the species. Many of the protected areas with rhinos have reached the limited number of individual they can support. This leads to human-rhino conflict as rhinos leave the boundaries of the protected areas to forage in the surrounding villages. Rhinos, mainly females, reportedly kill several people each year in India and Nepal.

Predation

Due to its large size, the Indian rhino's only real predator in the wild are large wild cats such as tigers that will prey on the Indian rhino calves and weak individuals. **Protection in the wild:** With strict protection from Indian and Nepalese wildlife authorities, greater one-horned rhino numbers have recovered from under 200 last century to around 3,333 today. However, poaching has remained high and the success is precarious without continued and increased support for conservation efforts in India and Nepal.

Measures taken for Conservation

The Indian and Nepalese governments have taken major steps towards Indian Rhinoceros conservation with the help of the World Wildlife Fund (WWF). The Kaziranga National Park and Manas National Park in Assam, Pobitora reserve forest in Assam (having the highest Indian rhino density in the world), Orang National park of Assam, Laokhowa reserve forest of Assam having a very small population and Royal Chitwan National Park in Nepal are homes for this animal. The Indian state of Assam is home to the largest population of greater-one horned rhinos, with more than 90 per cent in Kaziranga National Park.

World Wildlife Fund Programme: Translocation

WWF continues to play an active role in establishing new rhino populations by moving them from one protected area to another, a process called translocation. The rhinos are moved to protected areas within the Eastern Himalayas, particularly in India's Kaziranga and North Bank Landscapes. Several rhinos have already been moved from the Pabitora and Kaziranga reserves to Manas National Park, a World Heritage Site, in order to establish a new rhino population. WWF is a key partner to the Indian state government of Assam on this mission. In September 2012, a rhino monitoring team of Forest Department staff and WWF-India researchers were delighted to see a new born rhino calf trailing its mother in a remote section of Manas National Park in the northeast corner of India. Born to a rhino that was translocated in January 2012, the calf is a cause for celebration among conservationists. It indicates that translocated rhinos like this female are adapting well to their new home.

Tackling Illegal Wildlife Trade

WWF and TRAFFIC, the wildlife trade monitoring network, work to stop trafficking of rhino horn by funding antipoaching patrols and supporting intelligence networks in strategic locations to prevent rhinos from entering black markets in Asia. The South Asia Wildlife Enforcement Network (SAWEN) is now also working for this so that regional governments are able to combine information and resources. This includes early warning systems, investing in effective legislation, and improving enforcement of policies and laws.

Monitoring and Protection

WWF supports national park staff and rangers with equipment and training in order to protect rhinos from poaching and track other illegal activities such as logging. They also invest in improving rhino monitoring to collect data and measure progress towards achieving rhino conservation goals, assess the reproductive health and growth rate of populations, and make the right decisions to keep rhino numbers growing at a rate of more than 5 per cent.

Restoring Landscapes

As rhino populations increase, they need additional space to live and breed. WWF and partners restore rhino habitat in Nepal to increase rhino numbers and improve connectivity between protected areas. Chitwan National Park's population of greater one-horned rhinos is the second largest population of this species in the world, after India's Kaziranga National Park. In Kaziranga National Park, WWF works to secure habitat corridors so that rhinos have access to higher areas outside of the park during annual floods which inundate the park.

Working with Local Communities

The protected areas of India and Nepal, where rhinos reside, are surrounded by dense human populations. It is vital to ensure that communities that live around rhino reserves are sympathetic to, and benefit from, the rhinos in their midst. WWF supports several projects to improve local livelihoods, like the successful community-run "Marmelous juice factory" in Khata, Nepal. They help to decrease conflict between people and rhinos by encouraging farmers to plant unpalatable species like menthe (mint) that generates money for the communities and supports community-based antipoaching operations.

Strengthening Law Enforcement

WWF partners with national governments to strengthen wildlife laws and their enforcement, and fund antipoaching equipment and operations in protected areas. In response to a poaching spike in Nepal in 2006, WWF increased the number of security posts from eight to twenty. We also engaged ex-Army and police to patrol vulnerable points outside protected areas. Local youth volunteered to guard individual rhinos through the night. WWF relayed the information collected by these allies to key government departments so they could take action where needed. The number of rhinos poached in and around Chitwan in 2007 fell to only one. WWF expanded the operation to protect the diminishing rhino population in Bardia National Park, the second-largest stronghold for rhinos in Nepal. In 2011, Nepal celebrated a landmark year where zero rhinos were poached.

Conservation Challenges

For years, rhinos have been widely slaughtered for their horn, a prized ingredient in traditional Asian medicines. Destruction of their habitat over the years, has brought the rhinos to the brink of extinction. The great one-horned rhino could oncc bc found from Pakistan all the way through India, Nepal, Bangladesh, Bhutan and Myanmar. By the turn of the century, this species had vanished from much of its range, and today only about 3000 survive in India and Nepal. Throughout their range, their habitat continues to dwindle fast due to conversion of grassland habitats into agricultural fields and other human pressures. The threat of poaching continues to be ever-present.

WWF-India's Future Programme

Conserving the rhinos and their habitat is imperative. WWF has been working on rhino conservation for over four decades. The big programme initiated by WWF is the "Indian Rhino Vision 2020" (IRV 2020). The vision of the programme is to increase the total rhino population in Assam to about 3000 by the year 2020 and just as significantly ensure that these rhinos are distributed over at least seven protected areas to provide long-term viability of an Assam meta population of the species. This will be achieved by translocating the rhinos from two-source populations (Kaziranga and Pobitara) into 3 or 4 target Protected Areas (Manas, Laokhowa, Burachapori, Kochpora, Dibrusaikhowa and, possibly, Orang).

Other Challenges

The Forest Department faces a major challenge as lack of equipment, finance, political will and shortage of staff makes it difficult to implement conservation work at the grass root level. Two serious on the ground problems include, containing poaching and loss of habitat due to encroachments.

References

1. **Animal info: Indian Rhino** Biology, ecology, habitat, and status of the Indian Rhino and information on their native countries: biodiversity, ecosystems and population.[w] http://www.animalinfo.org/species/artiperi/rhinunic.htm.
2. http:/www.kaziranga-national park.com/one horned-rhino. shtml1#sthash.3FgcwNIj.dpufGreaterone- horned rhino information.
3. http://Indian per cent 20Rhino.htm.
4. **WWF: Indian Rhino Info** Extensive information by the World Wildlife Fund on the Indian Rhino. Covers physiology, threats, habitat and much more. http://www.panda.org/about_wwf/what_we_do/species/about_spec.dex.cfm.

Chapter 10

Himalayan Musk Deer (*Moschus chrysogaster*): An Introduction of the State Animal of Uttarakhand with Persisting Threats and Mitigation Strategies Employed

Amita Saxena and Pooja Kanyal

Department of Fisheries Resource Management
College of Fisheries, GBPUAT,
Pantnagar – 263 145, Uttarakhand

ABSTRACT

The word 'musk' is derived from the ancient Indian word 'muskah' meaning testicles. A musk gland or pod situated between the naval and genitals of the male secretes a brownish wax-like substance known as musk. Musk Deer (Moschus Chrysogaster) (Indian name: Kasturi Mriga) is a mammal of higher Himalaya hence it is also called Himalayan musk deer. It is state animal of Uttrakhand. Because of the indiscriminate killing of the Himalayan musk deer for its musk, this animal is on the verge of extinction. Himalayan musk deer is the least studied of the deer-like animals. Habitat destruction and predation pose threats to musk deer populations however large-scale of illegal hunting to meet commercial demand for the scent gland or 'pod'. It is protected by law in Bhutan, Nepal and India. In china hunting may be permitted in some areas, although a license is required. The uneven enforcement of legislation across its range has meant little impact on preventing the rampant trade of the species. Improving the enforcement of anti-poaching laws is the key for better conservation of this species. Kedarnath wildlife sanctuary also called as the Kedarnath Musk Deer Sanctuary is the national sanctuary in Uttarakhand, India, in addition to this a protected area is also situated in Uttarakhand.

Keywords: *Musk, Habitat, Scent gland, Anti-poaching laws.*

Listing of *Moschus* spp. (1996 IUCN Red List Category)

1. Alpine Musk Deer *Moschus sifanicus* LR (nt) I China and India
2. Black Musk Deer *Moschus fuscus* LR (nt) I II Bhutan, India, Myanmar and Nepal China
3. Forest Musk Deer *Moschus berezovskii* LR (nt) II China and Vietnam
4. Himalayan Musk Deer *Moschus chrysogaster* LR (nt) III Afghanistan, India, Nepal and Pakistan, China
5. Siberian or Taiga Musk Deer *Moschus moschiferus* VU II Russia, Kazakhstan, Kyrgyzstan, China, Korea and Mongolia

(*Source: Wemmer, 1998; Anon., 2003, IUCN, 2004*).

Physical Characteristics

Musk deer is generally nocturnal, shy and solitary. Groups usually include a mother and her offspring. They do not undertake any seasonal migration thereby remaining in the same area year-round despite harsh weather conditions (Kattel, 1992). The fawns are often spotted which gradually disappear early in life and adults have two white spots, one on either side of the neck below the jaws (Kattel, 1992). It has large ears and a short tail with long hairs at the arial region. Having dark brown color, its body is covered over with coarse and brittle hairs. Individual hairs contain air-filled cells for better insulation (Green, 1985). Musk deer movement appears more like jumping. Their toes are large for their body size and can be spread to find secure footing in mountains and on snow.

Habitat

Male musk deer are highly territorial. They tolerate only female musk deer within their home ranges and defending their home range against other males of the species, while female territories may overlap (Green, 1998). The territory of one male overlaps with parts of the territory of five females but not with parts of the territory of another male (Kattel, 1992).

Population and Distribution of Himalayan Musk Deer

The musk deer is distributed sporadically throughout the forested, mountainous part of Asia, ranging from just north of the Arctic Circle, south to the north-eastern edge of Mongolia and to Korea. It occurs further south (excluding the Gobi Desert) in China, northern Vietnam, Myanmar and the Hindu Kush-Himalayan region of Afghanistan, Pakistan and India. In Central Asia, musk deer occur in Kazakhstan, possibly in Kyrgyzstan, and the south of Russia (Dao, 1977; Flerov, 1952; Green, 1986 and Whitehead, 1972).

In India, it inhabits in parts of Kashmir, Himachal Pradesh, the northern part of Uttar Pradesh, Sikkim and Arunachal Pradesh. Its altitudinal range extends from about 2400 m to above the tree line, which varies from about 3200m in the Western Himalaya to 4200 m in the eastern Himalaya. The species is more common in the Eastern Himalayas (Green, 1985, 1987a).

Food Preference

Musk deer feeds on herbaceous and woody plants, leaves, flowers, twigs, lichen, moss, shoots and grass. More than 130 plant species are consumed by musk deer. In the winter, arboreal lichens and some terrestrial bushy lichens make up about 70 per cent of the contents of a musk deer's stomach (by weight).

Reproduction

The relatively high reproductive rate of musk deer has certainly contributed in preventing extinction of the species. The mating season is November-January, depending on area, altitude and region. Males mature at the age of 2.5-3 years and are capable of performing more than three copulations a day. Female musk deer mature at their first year (Green, 1987a and 1989). The birth weight of musk deer varies between about 400 g and 600 g, depending on the species. The young musk deer, in their first two months, are 'nursed offspring', hidden in the undergrowth and suckled by their mothers. At the age of about two months they start following their mothers and are weaned (Green, 1987a). The young grow rapidly, become independent of their mothers by the age of six months (Green and Kattel, 1997).

Communication between Animals

Communication among musk deer is mainly by olfaction, although they also have excellent hearing and vision. The role of musk in communication is not fully understood. Caudal (tail) and inter-digital glands also functions to leave scent-marks (Green, 1985). Red or pink-stained and sweet-smelling patches of urine observed on snow indicate that the musk of the male is probably emitted in the urine, while the urine of females is amber-colored and does not have any noticeable smell to humans (Green, 1987c). The scents could be used for territorial marking by animals.

The Musk Gland and Musk as a Product

Musk refers to the secretion from the scent gland of the male musk deer, the smell of which comes mainly from "muscone" – a component first isolated in 1906 (Homes, 1999). It is for their musk, above all else, that musk deer are hunted and trapped (Green and Kattel, 1997). Musk ripens in the musk sac within a month and gets converted to a scented red-brown substance. When removed from the animal, the gland is dried, whereupon the inner red-brown creamy substance blackens and becomes granular and powdery (Mukerji, 1953). Musk may get into the trade either as whole pods or as the granular red-brown contents of the pod, also known as musk grain. Male musk deer produce musk from the age of 12 to 18 months onwards; however, peak productivity occurs between the ages of 3–8 years old. During this period individuals produce an average of 25 grams of musk per year (Green, 1989). In captivity, males produce little musk after the age of 14 years but the ability to produce musk is retained until the age of 20 years. On average, 18 g (10 g dry weight) of musk can be harvested annually from animals kept in captivity (Green, 1989). Muscone, 3-Methylcyclopentadecanone, is the major constituent of natural musk. The odoriferous nucleus of natural musk is the C-16 cyclic ketone (cyclopentadecanone) (Anon, 2005a). Musk also contains macrocyclic compounds, steroids, proteins, esterols, esters, waxes, urogenic salts and the male hormone – androsterone (Tewari, 2001).

Uses of the Musk

a. As a Scent

It is used in perfume as a scent and fixative. It is obviously the first importance in perfumery as it features the better properties in not only penetration but also persistency than any other known substance. Synthetic musk should replace for natural musk to protect animals. Synthetic musk-like products are based on the symmetry of the three nitro groups or on polycyclic and cyclopentadecanone structure.

b. Therapeutic Uses

Musk carries medicinal value in Islamic countries, India and countries of the Far East (Green 1983). Its use in medicine was recorded by Aetius, the Greek physician, in circa 520 AD (Pereira, 1857, in Green 1983). In China and India musk has been considered a superior medicine since the fifth century AD (Green, 1983). However, as early as the Han Dynasty (2000 BC – 200 AD), records in Shennong Bencao Jing show the use of musk in traditional Chinese medicine. It is reported that musk is used in more than 300 traditional East Asian medicine preparations, being prescribed both as a sedative and a stimulant (Mills, 1998; Homes, 1999).

The south-east musk continues to be a popular traditional Chinese medicine for the effective treatment of improving blood circulation and relieving ailments of the heart, nerves and breathing system. It can revive unconscious patients, also possesses anti-inflammatory and analgesic effects. It is used in the treatment of delirium, stroke, unconsciousness, miscarriage, ejection of stillborn fetus, acute angina pectoris, acute abnormal pain, skin infection, sore throat, sprained joints, trauma, and paralysis. It occurs in three dominant forms, as oils or sprays, medicated plasters and raw musk powder (Ng and Burgess, 2004). The musk powder can be dissolved into water then used as a salve on parts of the body that suffer rheumatic pains. Musk powder can also be dissolved into wine or ginseng tea and consumed to improve the body's blood circulation. Preparation of musk can be used for both external and internal applications. Musk oils can be brought 'off the shelf' or blended with snake bile or various plant herbs. These are used to obtain immediate and short-term relief of rheumatic and muscular pains caused by blood stasis.

Musk Deer: Global Status

a. Musk Deer and CITES

Concern over the high levels of international trade in parts and products derived from musk deer saw all musk deer species *Moschus spp*. included in the Appendices of the CITES in 1979, with the purpose of regulating and monitoring control of international trade in musk deer, so that such trade did not threaten the survival of these species. Populations of musk deer in Afghanistan, Bhutan, India, Myanmar, Nepal and Pakistan were included in Appendix I in 1983, with the effect that international commercial trade in wild specimens from these populations was prohibited. All other musk deer species were listed in Appendix II, which means that international trade is allowed, but is strictly regulated, according to the provisions of the Convention.

b. Musk Deer and the Red List

All species of musk deer are listed in the IUCN Red List of Threatened Species (IUCN, 2004). The IUCN Red List is an inventory of the global conservation status of plants and animals, which uses a set of criteria to evaluate the extinction risk of thousands of species and subspecies. These criteria are relevant to all species and all regions of the world. There are nine Categories of Threat in the IUCN Red List system: Extinct, Extinct in the Wild, Critically Endangered, Endangered, Vulnerable, Near Threatened, Least Concern, Data Deficient and Not Evaluated. The Siberian Musk Deer is classified as Vulnerable, while the Himalayan Musk Deer, Forest Musk Deer and Black Musk Deer are all listed as Lower Risk/near threatened (a 1994 Red List Category).

Threats

1. Poaching and Trade

The main reason for the intense hunting of musk deer has always been the demand for musk. The small size of the musk gland means that it is easy to hide and transport, thus making detection of smuggling extremely difficult. Indiscriminate hunting methods such as traps mean that around three to five musk deer are killed for every male with a musk pod, weighing approximately 15-25 grams. While habitat destruction poses a threat to musk deer population, large-scale illegal hunting to fullfill the commercial demand for the scent gland or pod is believed to be responsible for its dramatic declines. Poachers use a number of techniques, ranging from guns to poison spears (Jackson, 1979).

2. Predators and Other Animals

M. chrysogaster has a shy nature (Sathyakumar, 1993) and is easily alarmed (Kattel, 1992). In the wild, musk deer are subjected to a number of natural predators, against which they defend themselves mainly by hiding or by their jumping gallop. The primary predator in the park of its range is the snow leopard (*Panthera uncia*). Other predators are wild cat (*Prionailurus bengalensis*), fox (*Vulpes vulpes*), wolf (*Canis lupus*), Lynx (*Lynx lynx*), wolverine (*Gulo gulo*), sable (*Martes zibellina*) and yellow-throated marten (*Martes flavigula*). In addition, the fawns are attacked by large birds of prey such as Large-billed crow (*Corvus macrohynchus*), and Upland Buzzard (*Buteo hemilasius*).

3. Grazing

Owing to high seasonality and low primary productivity, the Himalayan region supports relatively low ungulate/herbivore biomass. It is, therefore, obvious that with the increase in the biomass of domestic animals, wild ungulates such as musk deer suffered competitive exclusion.

Increased livestock grazing and associated impacts have led to low musk deer densities in many areas in Kedarnath Wildlife Sanctuary, India (Sathyakumar, 1993). Owing to the overexploitation of the pastures themselves, domestic animals enter the forest to compete for shrub and undergrowth with musk deer. Even where the canopy is intact, the dense under-story favored by musk deer for food and shelter is often extensively damaged by domestic livestock.

4. Timber and Firewood Collection

Throughout much of its range, the musk deer has been under pressure from increasing human populations. Subsistent forestry for local fuel and timber use occurs in sub-alpine and alpine regions, which is drastically destroying musk deer habitat. The use of timber in the building construction is high in the areas because constructing RCC buildings (need sand, cement, iron rods, etc.) is very difficult. Due to the cold temperature and high price of alternate energy source, the local people mainly use firewood as an energy source. The unmanaged firewood collection from the forest area may lead to the habitat destruction.

5. Habitat Destruction

Musk deer inhabit in forested mountainous regions which provide them both food and shelter and these habitats are still being converted for settlements, agriculture and so forth for the continually growing human population. In addition to the dramatic impacts that harvest for the trade in musk pods has had on musk deer populations over this century, the loss of suitable habitat has also been a significant factor (Green and Kattel, 1997) and in the longterm, habitat destruction may represent as serious a threat as poaching (Green 1986). Commercial forestry, tourism and erosion as a result of poor civil engineering also contribute to the negative impact on the forests (Green, 1986).

Others

6. Constructions

Musk deer habitats are still being converted for settlements, agriculture and so forth for the continually growing human population.

7. Tourism

Despite much advantages of tourism in the locality, one cannot ignore the negative effects of it on the conservation of biodiversity. Along with the income and development from the tourists, pollution, degradation, deforestation and loss of biodiversity have also been increasing.

8. Genetic Threat

The problem of maintaining genetic variability among small populations can be divided into two general concerns: the effect of loss of variation on the evolutionary stability of populations; and the effect of inbreeding, out breeding and heterosis on individual fitness. Genetic variability provides the raw material for natural selection, loss results in potentially adaptive traits and evolutionary options being lost (Franklin, 1980). As musk deer is distributed in smaller area in the World, this is a probable threat.

Management of Musk Deer

Management of our natural resources is essential to ensure that resources are managed sustainably and economically, environment remains healthy for future generations and can conserve, protect and enhance our biodiversity and ecological communities. Natural resource management, conservation and restoration are applications of ecological principles by humans. The key elements of management theme are the utilization of natural resources at a sustainable level with a focus on minimizing depletion and the reduction in input of pollutants to these resources. The focal points are income generation, conservation of biodiversity, combating desertification and conservation of mountain ecosystems. Special attention should be paid to the demand for natural resources generated by unsustainable consumption and to the efficient use of those resources consistent with the goal of minimizing depletion and reducing pollution. It was realized that mountains are important source of water, energy and biological diversity; these natural resources are often threatened by accelerated soil erosion, landslides and rapid loss of habitat and genetic diversity. Hence, the proper management of mountain resources and socio-economic development of the people deserve immediate action.

Musk Deer Farming

Farming or captive breeding of musk deer has been carried out in various parts of the world for decades. Parry-Jones and Wu (2001) have described captive breeding programmes in several Asian countries, most notably China, which has carried out such programmes since 1958. Many of the earlier attempts to farm musk deer in China failed because of high mortality rates. In addition the Government of Nepal has started a musk deer research facility at Godavari, Lalitpur since 1996.

Quantity of Musk

Assuming that 25 g of musk can be extracted per deer per year 175 g can be obtained in a sustainable manner. The quantity can be increased from the management of musk deer and its habitat.

Live Capture and Release

A potentially preferable capture technique for sustainable musk harvest is the drive-net technique developed by Kattel and Alldredge (1991). This is the technique Dr. Kattel and his village assistants used to conduct his study on musk deer. The only equipment this technique requires are several locally constructed collapsible nets, common veterinary tranquilizers, antibiotic cream and a stainless steel or silver scoop for musk extraction. In addition, 10–15 participants are needed to set up the nets and drive the musk deer into them. The nets are 6 m by 2 m when fully stretched out, have a 10 cm^2 mesh, and are locally woven from coconut fiber and cotton fiber ropes. The veterinary tranquilizers used are a combination of 2 per cent xylazine (Rompun) and 10 per cent ketamine hydrochloride. The best time to capture deer is either in the late morning or early afternoon while they are resting.

Laws Regarding Musk Deer

The resource system can be seen as nested within a set of external conditions comprising the legal and policy context for community based resource management. In order for communities to successfully carry out collective management their property rights or user rights, the resource should be recognized either by law or government policy. These rights, as well as structures and capacities needed to implement them, should also be supported by law, government policy, public institution like the jurisdiction system which can provide last resort dispute resolution mechanism, and by the policies of other influential actors (Pomeroy *et al.*, 1998). When assessing conditions, it is necessary to examine three different areas of external influence: international laws, national laws and policies and influential orders.

1. International Laws

Two international legal agreements directly relate to the conservation and exploitation of musk deer populations. The first is the CITES, which is an international law that regulates international trade in listed species and the second is the CBD, which is an international agreement to create and enforce specific types of domestic legislation designed to conserve biodiversity. India, as a signatory to both treaties and having ratified them, is responsible, under international law, for implementation and enforcing their provisions (Belbase, 1999; Homes, 1999). Both allow for support and implementation of *in situ* conservation strategies that uses financial benefits, gained through regulated exploitation, to act as incentives for both species and habitat conservation.

a. Convention on International Trade in Endangered Species (CITES)

CITES (1973) is meant to reduce the exploitation of vulnerable species and populations of flora and fauna by regulating trade in their products and constituents. The idea behind CITES is that regulation of international trade will reduce the financial incentive to over exploit species by limiting access to international markets. In order to accomplish this, the parties to the agreement designated three classes of species, listed in appendix to the convention, and three corresponding levels of regulation. Musk deer populations are listed under Appendix É or ÉÉ. Nepalese musk deer, along with Indian, Pakistani, and Afghan populations, have been listed under Appendix É since 1979 (CITES, 2002b). This means that the treaty does not permit any commercial trade in musk from musk deer resident in Nepal. However, the parties to the treaty have recognized that sustainable utilization of endangered species, including international trade in their products, may provide the financial incentives necessary to conserve their habitat and wild populations. To allow regulated trade in Appendix É listed species, for the benefit of their conservation, the conference of parties (COP) have created three categories of exemptions: population annotations, captive breeding programmes, and ranching programmes (IUCN, 1999). All three exemptions require that the production state apply to the COP for a down listing of the exploited population to Appendix ÉÉ. The producing state must be able to show that the population in question is able to withstand a specific level of exploitation. This means either a down listing of a specific wild population, as in the case of population annotations, or a down listing of the population contained in captive breeding or ranching programmes.

b. Convention on Biological Diversity (CBD)

The CBD is a less stringent and enforceable convention than CITES because it is an agreement that prescribe domestic measures a country must take to conserve biodiversity rather than international actions. The CBD recognizes both the national and international value of biological diversity and its objectives are to conserve biological diversity while promoting the sustainable use of its components and the fair and equitable sharing of the benefits arising out of this use (Balbase, 1999). Among the convention's many concerns its emphasis on *in situ* conservation is of particular importance to the issue of community based management of wild musk deer populations. The convention promotes *in situ* conservation, the protection of species in their place of origin, as the primary approach to biodiversity conservation because the maintenance and recovery of populations of species in their natural surroundings is deemed crucial. Included in this regard for *in situ* conservations is a recognized need to "provide the conditions needed for compatibility between present used and the conservation of biological diversity and the sustainable use of its components" (Balbase, 1999).

2. National Laws

The only legal document that directly pertains to musk deer was the National Parks and Wildlife Conservation Act of 1972 (HMGN, 1973). The Act recognized and protects musk deer in two ways: through listing *Moschus spp.* as a Schedule-1 protected species and by providing additional protection for populations located in protected areas designated under the authority of the Act.

a. Kedarnath Wildlife Sanctuary

Kedarnath wildlife sanctuary also called as the Kedarnath Musk Deer Sanctuary is the national sanctuary situated in Uttarakhand, India having primary aim to conserve the Himalayan Musk Deer. Consisting of an area of 975 km^2, it is the largest protected area in Western Himalayas. The animal is protected under the "Threatened Deer Programme" of the IUCN, with cooperation to the government of India and World Wide Fund for Nature. The sanctuary includes a breeding center at kharchaula kharak, both to help advance understanding of the animal conservation requirement and to breed it in captivity for the reintroduction to the wild.

b. Ascot Musk Deer Sanctuary

Located 54 km from Pithoragarh near Ascot in Uttarakhand, India. Primarily aimed for the conservation, rehabilitation and habitat restoration of musk deer.

Alternative Products of Musk

Some other animal and plant species are also associated with the same kind of odour as musk. They produce similar aromas or substances and some have similar properties despite different chemical structure.

1. The **Musk Rose (Pahadi gulaph):** ***Rosa moschata*** Rosaceae occurs in the Himalayas.
2. The **Jagat Pet:** ***Delphinium trichophorum*** Ranunculaceae occurs in the Himalayas. It can be used as an alternative medicine in place of musk in Tibet (Amje Sherpa, Namche, pers. comm., 2005).
3. The **Musk Mallow:** ***Hibiscus abelmoschus*** Malvaceae produces seeds that smell of musk and from which oil is obtained (Falbe and Regitz, 1995). Musk Mallow occurs in India and was introduced to the tropics. It is also used as a musk scent in the perfume industry.
4. The **Muskrat:** ***Ondatra zibethica*** was originally indigenous to North America and found in wetlands, rivers and lakes. The musk scent in the Muskrat is produced in two small glands between the anus and the genitals and is exuded when the animal is stimulated (Homes, 1999).
5. The **African Civet:** ***Civettictis civetta,*** occurring in sub-Saharan Africa and other African and Asian civet species are a valuable source of a musk-like substance called "civet" (Schreiber *et al.*, 1989). This yellowish secretion has the consistency of butter and is a product from scent glands located near the civet's anus.
6. The **Russian Desman:** ***Desmana moschata*** produces "musk" from glands at the base of its tail, which lends the animal a musky odour. In particular during the last century, the Russian Desman was hunted for its musk, for use in the perfume industry (Fons, 1988).
7. The **Musk Oxen:** ***Ovibos moschatus*** occur naturally in Greenland and north of the Arctic Circle in Canada and have been introduced into Norway, Sweden, Russia and the USA (Alaska). They do not, however, have a strong musky odour.
8. The **Suni:** ***Neotragus moschatus*** is an antelope inhabiting dry scrub land in south-east Africa. The Suni emits a strong musky odour from its pre-orbital glands (Walther, 1988).
9. The **Musk Milfoil:** ***Achillea erba-rotta*** sub-species ***moschata*** Asteraceae occurs in the central and southern Alps. It gives off a remarkable aromatic and spicy scent which has undoubtedly contributed to its name (Wendelberger, 1976).

References

Anonymous. 2003. *Checklist of CITES Species*. CITES Secretarial and UNEP–World Conservation Monitoring Centre, Geneva, Switzerland.

Anonymous. 2005a. *Cyclopentadecanolide.* http://www.chemicalland21.com. November 25, 2005.

Balbase, N. 1999. *National Implementation of the Convention on Biological Diversity: Policy and Legislative Requirement.* IUCN Nepal, Kathmandu, Nepal. Pp. 120.

CITES. 2002b. *Resolution of Conference 11.16: Ranching and Trade in Ranched Specimens of Species Transferred from Appendix I to Appendix II.* Convention on the International

Trade in Endangered Species. www.cites.org/eng/resols/11/11_16.shtml. November 25, 2005.

CITES. 2002c. *Resolution on Conference 11.7: Conservation of and Trade in Musk Deer.* Convention on the International Trade in Endangered Species. www.cites.org/eng/resols/11/11_7.shtml. November 25, 2005.

Dao, V.T. 1977. Sur Quelques Rares Mammifères au Nord du Vietnam. *Mitteilungen aus dem Zoologischen Museum Berlin* 53: 325–330.

Falbe, J. and Regitz, M. 1995. *Römpp Chemie Lexikon.* Thieme Verlag, Stuttgart, Germany and New York, USA.

Flerov, C. C. 1952. Fauna of the USSR, 1 (2). *Mammals: Musk Deer and Deer.* USSR Academy of Sciences, Moscow. (Translated from Russian by Israel Programme for Scientific Translations). pp. 14–45.

Fons, R. 1988. Heutige Insektenfresser. In: Grzimek, B. (Ed.). *Enzyklopädie Säugetiere.* 1: 425– 521. Kindler Verlag, Munich, Germany. 53

Franklin, I. R. 1980. Evolutionary Change in Small Populations. In: *Conservation Biology – an Evolutionary – Ecological Perspective.* Edited by M. E. Soule and B. A. Wileox. Sinauer Association. Sunderland, Mass. pp. 135–49.

Green, M. J. B and Kattel, B. 1997. *Musk Deer: Little Understood, Even its Scent.* Paper presented at The First International Symposium on Endangered Species Used in Traditional East Asian Medicine: Substitutes for Tiger Bone and Musk. 7–8 December 1997, Regal Riverside Hotel, Hong Kong. TRAFFIC/IUCN.

Green, M. J. B. 1983. The Musk Trade, with Particular Reference to its Impact on the Himalayan population of *Moschus chrysogaster*. Paper presented at the *Bombay Natural History Society Centenary Seminar on Conservation in Developing Countries,* Indian Institute of Technology, Powai, Bombay, 6–10 December 1983. pp. 36.

Green, M. J. B. 1985. *Aspects of the ecology of the Himalayan musk deer.* Ph.D. thesis, University of Cambridge, Cambridge, UK. pp. 280.

Green, M. J. B. 1986. *Distribution, Status and Conservation of Himalayan Musk Deer (Moschus chrysogaster).* Biological Conservation 35: 347–375.

Green, M. J. B. 1987a. *Some ecological aspects of a Himalayan population of musk deer.* In: Wemmer, C.M. (Ed.). The Biology and Management of the Cervidae. Smithsonian Institution Press, Washington, DC, USA.

Green, M. J. B. 1987c. *Scent–marking in the Himalayan musk deer (Moschus chrysogaster).* Journal of Zoology, London (B) 1: 721–737.

Green, M. J. B. 1989. *Musk Production from Musk Deer.* In: Hudson, R.J., Drew, K.R. and Baskin, L.M. (Eds.). Wildlife Production System. Cambridge University Press, Cambridge, UK.

Green, M. J. B. 1989. *Musk Production from Musk Deer.* Traffic Bulletin, 8: 59–61.

Green, M. J. B. 1998. *Musk deer: Little Understood, Even its Scent.* In: Proceedings of the First International Symposium on Endangered Species Used in Traditional

East Asian Medicine: Substitutes for Tiger Bone and Musk. Hong Kong, December 1997.

Homes, V. (Ed.) 1999. *On the Scent: Conserving Musk Deer – The Uses of Musk and Europe's Role in its Trade,* Brussels: TRAFFIC Europe. pp. 57.

IUCN, 1999. *Trade Measures in Environmental Agreements.* Economics, Trade and Environment Unit, UNDP. Gland, Switzerland.

IUCN, 2004. *IUCN Red List of Threatened Species.* www.redlist.org. July 03 2005.

Jackson, R. 1979. Aboriginal Hunting in West Nepal with Reference to Musk Deer *Moschus moschiferus moschiferus* and and Snow Leopard *Panthera unica. Biological Conservation.*16: 63–72.

Kattel, B. 1992. *Ecology of the Himalayan Musk deer in Sagarmatha National Park, Nepal.* Thesis Report Submitted to Colorado State University, USA.

Kattel, B. and Alldredge, A. 1991. Capturing and Handling of the Himalayan Musk Deer. *Wildlife Society Bulletin,* 19: 397–399.

Mills, J. 1998. *Need for Further Research into Tiger Bone and Musk Substitutes Agreed.* Trafic Dispatches, April 1998: 3–4.

Ng, D. and Burgess, E. A. 2004. *Against the Grain: Trade in Musk Deer Products in Singapore and Malaysia.* A TRAFFIC South Asia Report.

Parry–Jones, R., and Wu, J. 2001. *Musk Deer Farming and a Conservation Tool in China.* TRAFFIC East Asia, Hong Kong.

Pereira, J. 1857. *The Elements of Materia Medica and Therapeutics, 4th edn, II.* The Musk Trade, with Particular Reference to its Impact on the Himalayan Population of *Moschus chrysogaster.* Paper presented at the *Bombay Natural History Society Centenary Seminar on Conservation in Developing Countries,* Indian Institute of Technology, Powai, Bombay, December 6–10, 1983. pp. 36.

Pomeroy, R., Katon, B. and Harkes, I. 1998. *Fisheries Co–management: Key Conditions and Principals Drawn from Asian Experiences.* Presented at the 7th Biannual Meeting of the International Association for the Study of common Property. Vancouver, Canada.

Rajchal, R. 2006. Population Status, Distribution, Management, Threats and Mitigation Measures of Himalayan Musk Deer (*Moschus chrysogaster*) in Sagarmatha National Park. Report submitted to DNPWC/TRPAP.

Sathyakumar, S. and Prasad, S. N. 1990. *Reintroducing Captice Himalayan Musk Deer.* Zoo's Print VI: 7

Schreiber, A., Wirth, R., Riffel, M. and Rompaey, H. V. (1989). *Weasels, Civets, Mongooses, and their Relatives – An Action Plan for the Conservation of Mustelids and Viverrids.* IUCN/SSC Mustelid and Viverrid Specialist Group. IUCN, Gland, Switzerland.

Sherpa, L. G. 2005. *Sagarmatha National Park and Buffer Zone.* Sagarmatha Eco Kosheli, Sagarmatha Community Agro–Forestry Project., Namche, Solukhumbu. pp. 8–9.

Tiwari, I. 2001. Fading Fragrance of Musk. *Current Science*, Department of Forestry. Kumaun University, Nainital, India. 80: 12.

Walther, F.R. 1988. Ducker und Böckchen. In: Grzimek, B. (Ed.). *Enzyklopädie Säugetiere*. 5: 325– 343. Kindler Verlag, Munich, Germany.

Wemmer, C. 1998. *Deer Status Survey and Conservation Action Plan*. IUCN/SSC Deer Specialist Group. IUCN, Gland, Switzerland and Cambridge, UK.

Whitehead, G. K. 1972. *Deer of the World*. Constable and Company Ltd., London, UK.

Chapter 11

Biology of Sea Turtles and Need for Conservation

Bindu Sharma

Department of Zoology
D.N. College, Meerut, U.P.

ABSTRACT

There are seven known species of sea turtles that inhabit the world's oceans. They include the leatherback, green, olive ridley, Kemp's ridley, hawksbill, loggerhead and flatback turtle. Sea turtles have remarkable life cycles which makes them fascinating to both biologists and wildlife enthusiasts. They are found in all major ocean basins, and typically inhabit tropical and subtropical waters, although individuals are known to have been found in the colder waters of the Arctic. Once they leave the beach as hatchlings, females will return to land only to nest; male sea turtles never return to land at all. Most sea turtles undertake long distance migrations, as hatchlings and juveniles, during a long pelagic phase in the open ocean, and as adults between their feeding and breeding grounds. Loggerheads migrate over 12,000 km across the Pacific from their developmental habitats in Baja California to their nesting grounds in Japan. Green turtles migrate from their feeding grounds in Brazil to nest on Ascension Island, a speck in the middle of the Atlantic Ocean. On the east coast of India, olive ridley turtles that nest in Orissa are known to migrate to Sri Lanka and perhaps, beyond. Most species of sea turtles are listed as threatened. Worldwide, sea turtle populations face increasing threats that need to be addressed through appropriate research and conservation measures. There are a number of international, regional and national laws and regulations that offer protection to sea turtles; however, inadequate implementation and a growing conflict of interest with coastal development and expansion of the fishing industry have continued to take a toll on the species. Their global distribution and utilisation of diverse habitats makes sea turtles ideal flagships of the ocean, nearshore and coastal ecosystems.

Keywords: *Conservation, Habitat, Migration, Population, Sea turtles.*

Introduction

Sea turtles belong to class Reptilia and Order – Testudines. This order includes all turtles and tortoises. It is divided into three suborders: Pleurodira (side-necked turtles), Cryptodira (freshwater turtles, snapping turtles, tortoises, soft-shelled turtles, and sea turtles), and Amphichelydia (a suborder of turtles that is now extinct).Most scientists recognize two families of sea turtles: Family Cheloniidae includes all sea turtles with scutes (horny plates) covering their shells. Family Dermochelyidae are scuteless turtles with only one modern species; the leatherback turtle. A leatherback turtle is covered with leathery skin. It is the only marine turtle whose backbone is not attached to the inside of its shell.Most scientists recognize seven species and one subspecies of sea turtles:

Green sea turtle (*Chelonia mydas*); two subspecies the green (*Chelonia mydasmydas*) and the black or Eastern Pacific green turtle (*Chelonia mydasagassizii*)

Fossil Record

The first turtles appeared during the Triassic period, 245 to 209 million years ago.

The earliest known sea turtles appear in the fossil record in the Late Jurassic period, 208 to 144 million years ago. Scientists believe that modern sea turtles are derived from marsh-inhabiting ancestors that lived during the Late Triassic period.

Fossil records show that the now-extinct sea turtle Archelonischyros, which lived 144 to 65 million years ago, was one of the largest turtles that ever lived. It reached a length of 3–4 m (9.8–13 ft.).

Sea turtles, saltwater crocodiles, sea snakes, and marine iguanas are the only surviving reptiles that depend on the sea Sea turtle species are among a small group of reptiles that depend on the sea for survival.

Habitat and Distribution

Distribution

Various species of sea turtles are found in warm and temperate seas throughout the world.

Habitat

Adults of most species are found in shallow, coastal waters, bays, lagoons, and estuaries. Some also venture into the open sea. Juveniles of some species may be found in bays and estuaries, as well as at sea.

Sea Turtle Distribution in India

Five species of sea turtles, namely the olive ridley (*Lepidochely solivacea*), green (*Chelonia mydas*), leatherback (*Dermochelys coriacea*), hawksbill (*Eretmochelys imbricata*) and loggerhead (*Caretta caretta*), occur in the Indian coastal waters of the Bay of Bengal and Arabian Sea. All species except the loggerhead nest on the mainland coast and islands. The olive ridley nests on both coasts (primarily the east coast) and Andaman and Nicobar Islands. The green turtle nests in Gujarat, Lakshadweep and the Andaman and Nicobar Islands, the hawksbill in Lakshadweep and Andaman and Nicobar Islands and the leatherback in the Andaman and Nicobar Islands. In addition, practically all coastal states have sporadic nesting beaches and offshore congregation zones for sea turtles. Olive ridleys and leatherbacks, and hawksbills and greens often share nesting beaches in the Andaman and Nicobar Islands. Loggerheads are rare in India and known from very few records, mainly in southern Tamil Nadu. All five species nest in Sri Lanka.

Migration

Some sea turtle populations nest and feed in the same general areas; others migrate great distances.

Migration habits differ not only among species but also among different populations of the same species.

Green sea turtle populations migrate primarily along the coasts from nesting to feeding grounds. However, some populations travel 2,094 km (1,300 miles) across the Atlantic Ocean; from Ascension Island nesting grounds (in the middle of the South Atlantic) to Brazilian coast feeding grounds.

Loggerheads leave foraging areas and travel on breeding migrations that can be a few thousand kilometers (1 kilometer = 0.62 miles) each way.

Kemp's ridley turtles follow two major routes in the Gulf of Mexico: one northward to the Mississippi area, and the other southward to the Campeche Bank, near the Yucatan Peninsula.

Populations of olive ridleys have been observed in large flotillas traveling between feeding and nesting grounds in the Eastern Pacific and Indian Oceans.

Hawksbill migration studies have been limited. Evidence suggests that some hawksbill populations show cyclic nesting migrations. Other researchers have documented non-migratory and short-distance migratory populations.

Flatbacks move from their nesting grounds on the northern coast of Australia and its islands to feeding grounds in shallow waters of northeastern Australia. Distance covered ranges from 215–1,300 km (134–807 miles).

Leatherbacks have the longest migration of all sea turtles. They have been found more than 4,831 km (3,000 miles) from their nesting beaches.

Historically, scientists tracked a free-ranging sea turtle by tagging a flipper and documenting where the turtle was sighted. Although this method yields information on migration destinations, it does not reveal travel routes.

Hubbs-SeaWorld Research Institute has developed a radio/satellite transmitter harness for leatherback turtles. Its design allows secure attachment of a transmitter without affecting turtle mobility. The harness was designed to release within several months.

Population

Total population figures are often unknown because juvenile and male sea turtles do not come ashore and are difficult to count.

Population data are usually based on the numbers of adult females that come ashore to nest. Even then, the numbers are ambiguous – some females nest every two to three years; some may nest more than once on the same beach in a season; and some visit more than one nesting beach in a season.

Researchers rely on the changing numbers of nesting females from year to year to determine population trends. Because broad year-to-year fluctuations in numbers of nesting females make short-term data misleading, surveys of a decade or less may be insufficient to determine a population trend.

The Kemp's ridley is the most endangered sea turtle. In 1947, 42,000 nests were counted in a single day. The numbers declined dramatically until the 1980s. Currently, because of protection of nesting beaches and the use of turtle excluder devices in commercial shrimp fishing, Kemp's ridley populations seem to be slowly increasing.

Nesting populations of green sea turtles have not been surveyed long enough to determine worldwide population numbers. Experts estimate that in the past 120 to 140 years green sea turtle numbers have declined by 48 to 67 per cent (IUCN 2004). Between 200 and 1,000 green sea turtles nest on beaches in the continental U.S.

Little data are available on hawksbill populations. Probably about 15,000 female hawksbills nest each year worldwide, although accurate estimates of population sizes are difficult by aerial assessment. Tracks in the sand do not last long and are difficult to see, and nests are often obscured by beach vegetation.

The major loggerhead nesting grounds are in the southeastern US population trends of loggerheads show a decline in Georgia and South Carolina nesting areas, but no decline or a possible increase in southern Florida nesting areas. More years of nesting data and population biology studies are needed to assess the Florida trends. In the US, total estimates for loggerhead nests per year range from 68,000 to 90,000. The worldwide population is unknown, although most populations outside of U.S. waters are declining.

The olive ridley is the most abundant sea turtle in the world. There are probably more than several hundred thousand adult female olive ridleys. In 1991, an estimated 610,000 turtles nested in a single week on a beach in India.

The current population of flatback turtles is unknown. Because of its restricted distribution, the flatback is the most vulnerable of all sea turtles to habitat change or exploitation.

There are probably 26,000 to 43,000 adult female leatherbacks worldwide.

Sea Turtles Physical Characteristics

Size

Adult male and female sea turtles are equal in size.

Species	Distribution	Habitat
Green sea turtle (*Chelonia mydasmydas*)	Atlantic Ocean, Gulf of Mexico, along Argentine coast, Mediterranean Sea, and Indo-Pacific	Tropical and subtropical areas near continental coasts and around islands Green sea turtles are found in the Atlantic Ocean, Gulf of Mexico, Mediterranean Sea, and Indo-Pacific.
Black sea turtle (*Chelonia mydasagassizii*)	West coasts of North and South America, from central Baja California to Peru.	Bays and protected shores; not commonly seen in the open ocean Black sea turtles are found along the west coasts of North and South America, from Baja California to Peru.
Loggerhead sea turtle (*Caretta caretta*)	Worldwide	Coastal tropical and subtropical; ventures into temperate waters, to boundaries of warm currents; prefer coastal bays, but found in coastal streams, creeks, and open ocean. Loggerhead sea turtles are found worldwide in temperate and tropical habitats.
Kemp's ridley sea turtle (*Lepidochelys kempii*)	Adults usually occur in the Gulf of Mexico. Juveniles and immatures range between temperate and tropical coastal areas of the north-western Atlantic Ocean. Occasionally young turtles reach northern European waters and as far south as the Moroccan coast.	Shallow areas with sandy or muddy bottoms rich in crustaceans Adult Kemp's ridley sea turtles usually occur in the Gulf of Mexico.
Oliveridley sea turtle (*Lepidochelys olivacea*)	Tropical regions of the Atlantic, Indian, and Pacific Oceans (rarely to central California); nearly unknown around oceanic islands.	Mostly coastal; traveling or resting in surface waters Olive ridley sea turtles are found in the tropical regions of the Atlantic, Indian, and Pacific Oceans.
Hawksbill sea turtle (*Eretmochelys imbricata*)	Throughout central Atlantic and Indo-Pacific regions; most tropical of all sea turtles.	Near coral reefs and rocky outcroppings in shallow coastal areas Hawksbill sea turtles are found throughout central Atlantic and Indo-Pacific regions.
Flatback sea turtle (*Natator depressus*)	Indigenous to north-western, northern, and north-eastern regions of Australia; the most restricted range of all sea turtle species.	Completely coastal; does not go beyond the continental shelf Flatback sea turtles are found in the northern, north-western, and northeastern regions of Australia.
Leatherback sea turtle (*Dermochelys coriacea*)	The most widely distributed of all sea turtles; found in the Gulf of Alaska and south of the Bering Sea in the north-eastern Pacific to Chile in the south-eastern Pacific. In the Barents Sea, Newfoundland and Labrador in the North Atlantic; around Argentina and South Africa in the South Atlantic. Throughout the Indian Ocean; and to Tasmania and New Zealand in the southwestern Pacific. This species is found farther north than any other reptile, marine or terrestrial.	Highly oceanic, approach coastal waters only during breeding season Leatherback sea turtles are found in the north-eastern and south-eastern Pacific, in the North and South Atlantic, and throughout the Indian Ocean.

The leatherback is the largest of all living sea turtles. Mature leatherbacks reach about 1.2 to 1.9 m (3.9–6.2 ft.) and 200 to 506 kg (441–1,116 lb.). The largest leatherback recorded weighed 916 kg (2,019 lb.).

The Kemp's ridley and olive ridley are the smallest species, and reach about 55 to 75 cm (22–30 in.) and 30 to 50 kg (66–110 lb.).

Green sea turtles reach about 78 to 112 cm (31–44 in.) and 68 to 186 kg (150–410 lb.). The largest individual collected was 1.5 m (4.9 ft.) and 395 kg (871 lb.).

Loggerheads reach about 82 to 105 cm (32–41 in.) and 66 to 101 kg (146–223 lb.).

Hawksbills reach about 53 to 114 cm (21–45 in.) and 27 to 86 kg (60–190 lb.).

Flatbacks reach about 81 to 97 cm (32–38 in.) and 60 to 84 kg (132–185 lb.).

Body Shape

Sea turtles are characterized by a large, streamlined shell.

Coloration

Depending on the species, sea turtles range can be olive-green, yellow, greenish-brown, reddish-brown, or black in color.

The green sea turtle gets its name from the color of its body fat.

Flippers

Limbs are flippers adapted for swimming. Sea turtles are awkward and vulnerable on land.Forelimbs are long and paddle-like.Long digits are fused throughout the flipper.Only one or two claws are present on each fore flipper.A sea turtle swims with powerful wing-like beats of its fore flippers.Hind flippers serve as rudders, stabilizing and directing the animal as it swims. The hind flippers of some species are quite dexterous at digging nests in the sand.A sea turtle cannot retract its limbs under its shell as a land turtle can.

Head

A sea turtle cannot retract its head under its shell as a land turtle can.Sea turtles have large upper eyelids that protect their eyes.Sea turtles do not have an external ear opening. Researchers have found that sea turtles respond to low frequency sounds and vibrations.Sea turtles do not have an external ear opening. Like other turtles, sea turtles lack teeth. Jaw shape varies among species. Each species has a jaw shape adapted for its diet.

Shell

The dorsal (top) side of the shell is called the carapace.The top shell is called the carapace.

Depending on species, the adult carapace ranges in shape from oval to heart-shaped.In all species except the leatherback, the bony shell is composed of broadened, fused ribs, and the backbone is attached to the carapace.The bottom portion of the shell is called the plastron.In all species except the leatherback, the shell is covered with a layer of horny plates called scutesScutes are firm but flexible, not brittle.Scientists can identify sea turtle species by the number and pattern of scutes.The leatherback turtle has a thick and oil-suffused skin, which is an excellent insulator, allowing this species to venture into cold water.The leatherback's carapace is composed largely of cartilage raised into prominent longitudinal ridges. A layer of thousands of small dermal bones lies just below the leathery skin.

A sea turtle's large, bony shell provides protection from predation and abrasion.

Sexual Dimorphism

Male and female sea turtles do not differ externally until they approach maturity.

Adult males have longer, thicker tails than females, because the male reproductive organ is housed in the base of the tail. In males, the tail may extend beyond the hind flippers.With the exception of the leatherback turtle, the claws on the fore flippers of sea turtle males are elongated and curved, which may help them grasp a female's shell during mating.

Senses

Hearing

All reptiles, including sea turtles, have a single bone in the middle ear that conducts vibrations to the inner ear. Researchers have found that sea turtles respond to low frequency sounds and vibrations.

Eyesight

Sea turtles can see well under water but are shortsighted in the air.Under experimental conditions, loggerhead and green sea turtle hatchlings exhibited a preference for near-ultraviolet, violet, and blue-green light.

Tactile

A sea turtle is sensitive to touch on the soft parts of its flippers and on its shell.

Taste

Little is known about a sea turtle's sense of taste.

Smell

Most researchers believe that sea turtles have an acute sense of smell in the water. Experiments show that hatchlings react to the scent of shrimp. This adaptation helps sea turtles to locate food in murky water.A sea turtle opens its mouth slightly and draws in water through the nose. It then immediately empties the water out again through the mouth. Pulsating movements of the throat are thought to be associated with smelling.

Adaptations for an Aquatic Environment

Swimming

Sea turtles are strong swimmers. The cruising speed for green sea turtles is about 1.5 to 2.3 kph (0.9–1.4 mph). Leatherbacks have been recorded at speeds of 1.5 to 9.3 kph (0.9–5.8 mph).

Forelimbs are modified into long, paddle-like flippers for swimming.

Neck and limbs are nonretractile. The shell adaptations necessary for retractile limbs would impede rapid swimming.

With the exception of females that come ashore to nest, sea turtles spend their entire lives at sea and are well adapted to an aquatic existence.With the exception of females that come ashore to nest, sea turtles spend their entire lives at sea and are well adapted to an aquatic existence.

Diving

Sea turtles are excellent divers. Leatherbacks routinely dive more than 305 m (1,000 ft.). They may reach depths of more than 1,190 m (3,900 ft.) seeking jellyfish for prey.Since they are cold-blooded, sea turtles have a slow metabolic rate. This slowed metabolism allows them to stay submerged for long periods of time.

Hawksbill turtles have been known to remain submerged for 35 to 45 minutes.

Green sea turtles can stay under water for as long as five hours. Their heart rate slows to conserve oxygen: nine minutes may elapse between heartbeats.

In the north-central Gulf of California, black sea turtles return each year to specific areas. They bury themselves in sand or mud under water and may remain dormant from November to March.

Although they can remain submerged for long periods, sea turtles must return to the surface for a breath of air.

During long dives, blood is shunted away from tissues tolerant of low oxygen levels towards the heart, brain, and central nervous system.

Leatherbacks have high concentrations of red blood cells; therefore, their blood retains more oxygen. The muscle of leatherbacks has a high content of the oxygen-binding protein myoglobin. Myoglobin transports and stores oxygen in muscle tissue.

Respiration

In studies conducted on green sea turtles, lung capacity exchange in one breath exceeded 50 per cent.

Salt Secretion

Sea turtles can live in sea water with no need for a freshwater source. They obtain sufficient water from their diet and from metabolizing sea water.

Like other marine reptiles and seabirds, sea turtles have a salt gland to rid their bodies of excess salt. This gland empties into the sea turtles' eyes. The secretion of salt and fluid makes them look as if they are "crying" when they come ashore. These "tears" also help keep the eyes free of sand while females dig their nests.

Sea Turtles on Land

For the most part, the only time sea turtles leave the sea is when females haul out to lay eggs. On some uninhabited or sparsely-inhabited beaches, turtles have been observed basking on land.

Many adaptations that make sea turtles successful in the sea make them slow and vulnerable on land.

Sea turtles are slow and vulnerable on the land.

Behavior

Social Behavior

Sea turtles are not generally considered social animals; however, some species do congregate offshore.

Sea turtles gather together to mate. Members of some species travel together to nesting grounds.

After hatchlings reach the water they generally remain solitary until they mate.

Individual Behavior

Little is known about the individual behavior of sea turtle species.

In the ocean, flatback turtles may spend hours at the surface floating, apparently asleep or basking in the sun. Frequently, seabirds perch on the backs of the flatbacks.

Hawksbill turtles spend some time resting or sleeping wedged into coral or rock ledges.

Olive ridleys have been observed basking on beaches, and it is not unusual to see thousands of olive ridleys floating in front of the nesting beaches.

Leatherback turtles tend to dive in a cycle that follows the daily rising and sinking of the dense layer of plankton and jellyfish. The turtles probably feed in the upper layers of water at night. As dawn approaches, their dives become deeper as the plankton and jellyfish retreat to deeper water, away from the light of day. The turtles bask at the surface at midday when the layer sinks beyond their typical diving range. As dusk approaches, the turtles' dives become shallower as the layer rises.

Green sea turtles are considered solitary, but occasionally form feeding aggregations in shallow waters abundant in seagrass or algae.

Diet and Eating Habits

Food Preferences and Resources

Depending on the species, sea turtles may be carnivorous (meat eating), herbivorous (plant eating), or omnivorous (eating both meat and plants). The jaw structure of many species is adapted for their diet.

Green sea turtles have finely serrated jaws adapted for a vegetarian diet of sea grasses and algae. As adults, these are the only herbivorous sea turtles.

Some species change eating habits as they age. For example, green sea turtles are mainly carnivorous from hatching until juvenile size; they then progressively shift to an herbivorous diet.

A hawksbill has a narrow head with jaws meeting at an acute angle, adapted for getting food from crevices in coral reefs. They eat sponges, tunicates, shrimps, and squids.

Loggerheads' and ridleys' jaws are adapted for crushing and grinding. Their diet consists primarily of crabs, molluscs, shrimps, jellyfish, and vegetation.

Leatherbacks have delicate scissor-like jaws that would be damaged by anything other than their normal diet of jellyfish, tunicates, and other soft-bodied animals. The mouth cavity and throat are lined with papillae (spine-like projections) pointed backward to help them swallow soft foods.

Researchers continue to study the feeding habits of flatbacks. There is evidence that they are opportunistic feeders that eat seaweeds, cuttlefish, and sea cucumbers.

In a zoological environment all sea turtle species can be maintained on a carnivorous diet.

While their diets may vary in the wild, in a zoological environment all sea turtle species can be maintained on a carefully managed carnivorous diet.While their diets may vary in the wild, in a zoological environment all sea turtle species can be maintained on a carefully managed carnivorous diet.

Reproduction

Sexual Maturity

Estimates of sexual maturity in sea turtles vary not only among species, but also among different populations of the same species. Maturity may range from as early as three years in hawksbills to 12–30 years in loggerheads to 20–50 years in green sea turtles.

Sexual maturity often is related to carapace size. Studies have shown that hawksbills reach sexual maturity at a carapace size of 60 to 95 cm (24–37 in.); loggerheads reach maturity at a carapace size of 79 cm (31 in.); and green sea turtles reach maturity at 69 to 79 cm (27–31 in.).

Evidence suggests that some turtles continue to grow after reaching sexual maturity, while some stop growing after reaching maturity.

Mating Activity

For most species, courtship activity usually occurs several weeks before the nesting season.

Two or more males may court a single female.

Male sea turtles of all species except leatherbacks have enlarged claws on their front flippers. These help grasp the shells of the females during mating.

Fertilization is internal. Copulation takes place in the water, just offshore.

Nesting Behavior

Like other turtles, sea turtles lay eggs.

Females come ashore on a sandy beach to nest a few weeks after mating.

Females usually nest during the warmest months of the year. The exception is the leatherback turtle, which nests in fall and winter.

Females of most species usually come ashore at night, alone, most often during high tide. A female sea turtle crawls above the high tide line and, using her front flippers, digs out a "body pit". Then using her hind flippers, she digs an egg cavity. The depth of the cavity is determined by the length of the stretched hind flipper.

A female deposits 50 to 200 (depending on the species) Ping Pong ball-shaped eggs into the egg cavity. The eggs are soft-shelled, and are papery to leathery in texture. They do not break when they fall into the egg cavity. The eggs are surrounded by a thick, clear mucus.

Sea turtle eggs are soft-shelled and papery to leathery in texture. A female may deposit as many as 50 to 200 (depending upon species) Ping Pong ball-shaped eggs into the egg cavity that she digs in the sand.

The female covers the nest with sand using her hind flippers. Burying the eggs serves three purposes: it helps protect the eggs from surface predators; it helps keep the soft, porous shells moist, thus protecting them from drying out; and it helps the eggs maintain proper temperature. Experts can identify the species of turtle by the type of mound left by the nesting female and by her flipper tracks in the sand.

Females may spend two or more hours out of the water during the entire nesting process.

It is possible that through the storage of sperm from one or several males in the oviducts of the females, all clutches of the current nesting season may be fertilized without repeated mating.

Female Kemp's ridley and olive ridley sea turtles form masses called arribadas (Spanish for "arrival"). Arribadas contain thousands of egg-bearing females that come ashore at the same time to lay eggs.

Most females return to the same beach where they hatched to nest each year.

Recent studies suggest that some females of some species will visit more than one nesting beach (other than the original beach) in a season.

Females usually lay between one and nine clutches (groups) of eggs per season.

Females may nest every two or three years.

Hatching and Care of Young

Incubation

Incubation time varies with species, clutch size, and temperature and humidity in the nest.

The incubation time for most species is 45 to 70 days.

Research indicates that the sex of an embryo is determined sometime after fertilization, as the embryo develops, and may be temperature dependent. Lower nest temperatures produce more males; higher temperatures produce more females.

Hatching

Sea turtles hatch throughout the year but mostly in summer.

Hatchlings use a carbuncle (temporary egg tooth) to help break open the shell.

After hatching, the young turtles may take three to seven days to dig their way to the surface.

Hatchlings usually wait until night to emerge from the nest. Emerging at night reduces exposure to daytime predators. They leave the nest and head to the water in groups. Studies have shown that some nests will produce hatchlings on more than one night.

Reaching the Ocean

There are several theories as to how hatchlings find the sea.

Hatchlings may distinguish light intensities and head for the greater light intensity of the open horizon.

During the crawl to the sea, the hatchling may set an internal magnetic compass, which it uses for navigation away from the beach.

When a hatchling reaches the surf, it dives into a wave and rides the undertow out to sea.

A "swim frenzy" of continuous swimming takes place for about 24 to 48 hours after the hatchling enters the water.

This frantic activity gets the young turtle into deeper water, where it is less vulnerable to predators.

There have been reports of swimming hatchlings diving straight down when birds and even airplanes appear overhead. This diving behavior may be a behavioral adaptation for avoiding predation by birds.

Upon reaching the surf, sea turtle hatchlings begin a "swim frenzy" of continuous swimming for about 24 to 48 hours after entering the water.

The First Year

During the first year, many species of sea turtles are rarely seen. This first year is known as the "lost year".Where hatchlings go and how they survive is still a mystery that researchers continue to study.

Researchers generally agree that most hatchlings spend their first few years living an oceanic existence before appearing in coastal areas. Although the migratory patterns of the young turtles during the first year has long been a puzzle, most researchers believe that they ride prevailing surface currents, situating themselves in floating seaweed where they are can find food.

Research suggests that flatback hatchlings do not go through an oceanic phase. Evidence shows that the young turtles remain inshore following the initial swim frenzy. Most remain within 15 km (9.3 miles) of land.

Sea Turtles Longevity and Causes of Death

Longevity

Scientists are still researching sea turtle longevity. Once sea turtles reach sexual maturity, they may have an estimated reproductive life of about 30 years. Given that some species reach maturity at 50 years, an 80-year life span is feasible.

Aging

Currently there is not an adequate method of aging sea turtles. The most accepted method is to study growth rings of the scales on the carapace and plastron. Scientists count the rings and use a mathematical formula to estimate a turtle's age.

Natural Predators

Adult sea turtles have a few predators, mostly large sharks. Tiger sharks, in particular, are known for eating sea turtles. Killer whales have been known to prey on leatherback turtles.

Fishes, dogs, seabirds, raccoons, ghost crabs, and other predators prey on eggs and hatchlings. More than 90 per cent of hatchlings are eaten by predators.

Flatback turtle nests are susceptible to predation by monitor lizards, dingoes, and introduced foxes.

Fibropapillomas

Green sea turtles may develop fibropapillomas (lobed tumor-like growths) on the skin. These growths can reduce vision, obstruct normal swimming and feeding, and increase susceptibility to secondary parasitism and infection.

Fibropapillomas on the skin of some green sea turtles and black sea turtles can reduce vision, obstruct normal swimming, feeding and increase the susceptibility to secondary parasitism.

Natural Disasters

Hurricanes can severely impact sea turtle nesting success. Almost 90 per cent of sea turtle nesting in the continental US occurs on Florida's beaches. Scientists estimate that more than half of the sea turtle nests on the coast of Florida were washed away by the four hurricanes hitting the coast in 2004.

During a severe cold snap in 1989, dozens of hypothermic green sea turtles were rescued and cared for in recovery pools at SeaWorld Orlando for about 10 weeks.

Human Impact

Sea turtle nesting areas are becoming scarce due to beach development and disturbances. Kemp's ridleys mainly nest on a remote beach in Mexico near the village of Rancho Nuevo (about 161 km, or 100 miles, south of the Texas border). In 1947, scientists witnessed an arribada of more than 42,000 Kemp's ridley turtles in one day. In the 1960s numbers were reduced to less than 5,000 turtles. In 1973, the largest arribada contained only 200 individuals. More recently, a small but growing number of Kemp's ridleys have nested on some beaches along the coast of Texas.

Nesting females and hatchlings are disturbed by the presence of trash on nesting beaches. If trash impedes its crawl up the beach, a female returns to the sea instead of nesting.

The noise and activity of people on the beach also may cause females to return to the sea instead of nesting.

Some sea turtles die when they ingest trash. Leatherbacks are especially susceptible to ingesting plastic, mistaking it for jellyfish.

Thousands of sea turtles drown in commercial shrimp nets each year. Shrimpers accidentally entangle turtles that are foraging where trawlers are working. This mortality may be increasing. In 1947, 5,000 US shrimping trawlers worked in the Gulf of Mexico. That number increased to 15,000 full-time and 40,000 part-time trawlers by 1989.

Artificial lighting on beaches may misrepresent the time of day to turtles attempting to nest. Most turtles are nocturnal nesters, and to a turtle that has not yet come ashore to nest, a brightly lighted beach may signify daylight and inhibit nesting.

Hatchlings can become disoriented by city and street lights when trying to find the surf. Many young turtles actually head away from the ocean and towards parking lots. These animals may be eaten by predators or crushed by cars. Some die from exposure.

Some people illegally collect turtle eggs for food and for their alleged aphrodisiac effect.

Sea turtles are hunted (illegally in this country and, in some cases, legally elsewhere) for their meat and shells, which are used to make combs, eyeglass frames, aphrodisiacs, and curios. The fat of green sea turtles, boiled with cartilage, made a popular soup called calipee, which led to the decline in green sea turtle population numbers.

Deforestation may indirectly threaten sea turtle nests. Costa Rica has one of the highest deforestation rates in the world. Some researchers fear that without the forest to draw up ground water, the water table will rise beneath beaches and drown nests.

Propeller and collision injuries from boats are not uncommon. These types of injuries are more frequent in areas with a high level of recreational boating, such as southern Florida, the Florida Keys, and the United States Virgin Islands.

SeaWorld veterinarians treat a sea turtle with damage to its shell.

Conservation and Research

Global Status

The world's seven sea turtle species are classified as follows according to the IUCN Red List of Threatened Species:

- ☆ Leatherbacks: Vulnerable
- ☆ Greens: Endangered
- ☆ Loggerheads: Endangered
- ☆ Hawksbills: Critically Endangered
- ☆ Olive Ridleys: Vulnerable
- ☆ Kemp's Ridleys: Critically Endangered
- ☆ Flatbacks: Data Deficient

IUCN Definitions

- ☆ Critically Endangered: Species is considered to be facing an extremely high risk of extinction in the wild.
- ☆ Endangered: Species is considered to be facing a very high risk of extinction in the wild.
- ☆ Vulnerable: Species is considered to be facing a high risk of extinction in the wild.
- ☆ Data Deficient: Species that cannot be evaluated because of insufficient information.

Legal Protection for Sea Turtles

The Endangered Species Act, 1973 (ESA)

The Endangered Species Act of 1973 (ESA) is administered by the US Departments of Interior and Commerce. It seeks to stop the extinction of wild animals and plants in the United States, other nations, and at sea. All sea turtles except the flatback are listed as threatened or endangered on the US endangered and Threatened Wildlife and Plants List. It is illegal to harm, or in any way interfere with, a sea turtle or its eggs.

Under the ESA, the hawksbill, Kemp's ridley, leatherback, populations of green sea turtles (along Florida and the Pacific coast of Mexico), and breeding populations of olive ridleys (on the Pacific coast of Mexico) are listed as "endangered" (species face a very high risk of extinction).

The loggerhead, green (except the populations listed above), and olive ridley (except the populations listed above) sea turtles are listed as "threatened" (species face a high risk of extinction).

Convention on International Trade in Endangered Species of Wild Fauna and Flora (CITES)

CITES is an international treaty, developed in 1973, to regulate trade in certain wildlife species.

CITES Appendix I includes species identified as endangered (species faces a very high risk of extinction). All sea turtle species are listed under Appendix I in CITES.

IUCN/The World Conservation Union is a Worldwide Conservation Organization

This organization links together government agencies, non-government agencies, and independent states to encourage a worldwide approach to conservation.

The hawksbill and Kemp's ridley sea turtles are both listed as critically endangered (the species faces an extremely high risk of extinction) by the IUCN.

The green, loggerhead, and olive ridley sea turtles are all listed as endangered (species faces a very high risk of extinction).

Turtle Excluder Device

At a cost of millions of dollars, the US National Marine Fisheries Service developed the Turtle Excluder Device (TED).

The TED is a small, metal grid trapdoor inside a trawling net that allows shrimp to pass to the back while the turtles escape to safety before becoming entrapped or entangled.

Since 1989, federal law requires that this device be installed on the nets of all US fishing trawlers working in areas populated by sea turtles.

Protecting Nests

Sometimes field biologists protect sea turtle nestlings from predators by placing screens over them. They may also relocate eggs laid too close to the water or in erosion zones to safer areas.

Sea turtles and their nests are protected by the Endangered Species Act and the Florida Marine Protection Act.

In a bold conservation programme, the towns people in one small Costa Rican beach village are allowed to gather eggs during the first two nights of each olive ridleyarribada. Scientists have calculated that a controlled harvest would leave enough protected eggs to rejuvenate the population (in one nesting season, 20 to 30 million olive ridley eggs may be laid in this beach village) while allowing villagers to maintain a livelihood. The programme has the potential to stop egg poachers on other beaches by keeping the prices of the "legal" eggs too low for poachers to compete.

Lighting

Although eliminating beach lighting would be the most effective way to reduce disorientation of hatchlings, studies have shown that low pressure sodium vapor lights have a lesser effect on loggerhead and green sea turtle hatchlings. Many beach communities have encouraged the use of these lights.

Wildlife Refuges

In 1991, Congress established the Archie Carr National Wildlife Refuge on the east coast of Florida. After completion of the acquisition of privately-owned, mostly undeveloped lands, the refuge will cover 900 acres, including 33 km (20.5 miles) of important sea turtle nesting habitat between Melbourne Beach and Wabasso Beach. Federal, state, and local agencies and private groups are sharing the projected $80 million cost to obtain the lands to complete the reserve.

This 33 km (20.5 mile) section of beach is the most important nesting site for loggerheads in the Western Hemisphere.

The refuge is also the most important nesting beach in the United States for the green sea turtle.

The refuge is the northernmost point on the Atlantic coast for leatherback nesting – in 1996 only ten nests were counted in or just adjacent to the refuge, in 2001 the number of nests counted steadily increased to 39.

The refuge is also considered prime real estate for commercial development, making government funding essential to its preservation.

The governments of Nicaragua and Costa Rica have established, and are striving to expand, national parks and biological reserves where sea turtles forage and nest. Tortuguero, Costa Rica maintains the largest green sea turtle rookery in the Caribbean. Local economics is no longer based on turtle harvests, but on tourism. More than 15,000 visitors are expected each year.

The Archie Carr National Wildlife Refuge provides protected nesting beaches for sea turtles and is an especially critical habitat for green sea turtles and loggerhead sea turtles that use this area to nest in large numbers.

Managing Sex Ratios

Most conservationists believe that abundant nesting females are desirable to rejuvenate sea turtle populations. Researchers with Reptile Conservation International have developed and patented a technique of applying an estrogen solution onto eggs to produce a higher number of females under normal incubation.

In Zoological Environments

Having sea turtles at marine zoological parks provides an opportunity for the public to learn, up-close, about these animals and how human activities may impact their survival. SeaWorld San Diego, SeaWorld Orlando, SeaWorld San Antonio, and Busch Gardens Tampa all offer a Saving a Species Tour in which guests can learn about endangered animals including sea turtles. A portion of the tour proceeds goes to the SeaWorld-Busch Gardens Conservation Fund – a non-profit charitable organization that sponsors sea turtle research along with other projects.

Sea turtles at marine zoological parks provide an opportunity for the public to learn about these animals and how human activities may impact their survival.Sea turtles at marine zoological parks provide an opportunity for the public to learn about these animals and how human activities may impact their survival.

In the protected environment of a marine zoological park, scientists can examine aspects of sea turtle biology that are difficult or impossible to study in the wild.

Project: Loggerhead Turtle Tracking

Partner: National Oceanic and Atmospheric Administration

Location: Cape Verde Archipelago, Western Africa

The Cape Verde archipelago represents one of the largest loggerhead nesting populations in the Atlantic Ocean and is likely the largest nesting population in western Africa. Although some areas of the archipelago are protected, human predation of sea turtles is significant. Satellite tracking technology allows NOAA to track turtles from Cape Verde to identify important life history information, migratory pathways, and feeding grounds. This information will provide the support required for increased protection of the loggerhead turtle. The SWBGCF grant allowed NOAA to purchase satellite transmitters needed to conduct this study.

Project: Environmental Co-factors and Fibropapillomatosis in the Green Sea Turtle

Partner: University of Central Florida

Location: Florida, USA

Fibropapillomatosis (FP) is an infectious disease threatening endangered Florida populations of green sea turtles (Cheloniamydas). Fibropapillomatosis is prevalent in degraded marine environments, but the environmental cofactors that increase incidence of FP remain unknown. The University of Central Florida study uses novel bio-molecular technology – a cellular-diagnostic system (CDS) – to identify environmental stressors – associated with FP among 3 populations with differing occurrences of the disease. The SWBGCF grant partially funds the study.

Project: Reducing Marine Turtle Bycatch in the Fisheries of the Eastern Pacific Ocean through Improvement in Fishing Gear and Techniques

Partner: World Wildlife Fund

Location: Eastern Pacific Ocean

The future survival of leatherback and loggerhead sea turtles is threatened by fisheries bycatch, particularly bycatch in longline and coastal gill net fisheries.

Fortunately, advances in fishing equipment and methods could dramatically reduce this bycatch. Recent research has demonstrated that turtle bycatch can be reduced by as much as 90 per cent through changes to longline fishing gear and techniques, including the use of circle hooks instead of traditional "J" hooks. Acting on these findings, WWF is promoting widespread improvement in fishing gear and techniques by engaging industry and governments in testing and promoting circle hooks and other improvements, turning a lose-lose situation into a win-win. WWF is also exploring ways in which gill net fishing might be improved. The SWBGCF grant helps support this industry and government outreach campaign.

Project: Movements and Habitat Preference of the Kemp's Ridley Sea Turtle

Partner: Texas A and M University of Galveston

Location: Northwestern Gulf of Mexico

The critically endangered Kemp's ridley sea turtle is experiencing a modest recovery from less than 300 nesting females in 1985. Continued species recovery and eventual downlisting to threatened status require researchers to update the Kemp's Ridley Recovery Plan, currently lacking information about the ocean life of this species. This project helps rescue, rehabilitate and track Kemp's ridley sea turtles accidentally caught by recreational fishermen. The satellite-tagged animals are tracked to monitor post-release behavior and survival rates. Data collected from this study helps researchers understand the turtle's use of near shore Gulf waters, habitat preferences, short-term movements, long-term migratory behavior and potential for exposure to fisheries interactions. The SWBGCF grant helps ensure a continuance of the western Gulfs only in-water monitoring programme for Kemp's ridley sea turtles.

Project: Abatement of Egg Poaching in a Pacific Leatherback Nesting Site

Partner: World Wildlife Fund

Location: Costa Rica

This project aims to increase the percentage of hatchlings from leatherback and black sea turtle nests occurring on an unprotected Costa Rica beach. Pacific leatherbacks are at risk of extinction unless drastic measures are taken. Every egg counts. Poaching currently impacts two-thirds of leatherback nests and all black turtle nests at Playa Junquillal, a recently discovered beach with a high number of nests. Increasing awareness of the benefits of marine turtle conservation among nest poachers and egg consumers, this project promotes alternative income sources dependent on live marine turtles. The SWBGCF grant helps fund a comprehensive community education programme encouraging poachers to generate income by participating in research and conservation programmes.

The SeaWorld and Busch Gardens Conservation Fund works on behalf of wildlife and habitats worldwide with the goal of encouraging sustainable solutions through the support of critical conservation initiatives around the globe.

The SeaWorld and Busch Gardens Conservation Fund works on behalf of wildlife and habitats worldwide with the goal of encouraging sustainable solutions through the support of critical conservation initiatives around the globe.

Conservation of Sea Turtle in India

Threats and mitigation

Estimating Threats to Sea Turtle Populations and Habitats

In many parts of the world and in India, sea turtle populations are affected by a wide variety of threats. Even under natural conditions, survival rates are low, and eggs and hatchlings are predated by small carnivorous mammals, birds, lizards and crabs. Once they are in the sea, a variety of predators plague them through their immature stages. Only large sharks, and perhaps killer whales, predate adults. In some locations, humans also catch and consume adults. At a few sites, nesting turtles may be killed by large predators such as jaguars and tigers. Human induced threats are increasingly problematic for turtle populations. These threats can be broadly classified into direct and indirect threats.

Direct threats include:

- ✰ Incidental catch in mechanised fisheries
- ✰ Consumption of adults – not very common in most parts of the Indian coastline
- ✰ Egg depredation by feral animals and humans

Indirect threats include:

- ✰ Loss of marine habitats (due to pollution, aquaculture, coastal tourism, etc.)
- ✰ Loss of nesting beaches (due to erosion, sand mining, beach armouring, pollution, exotic plantations, etc.)
- ✰ Artificial lighting (disorientation of both adults and hatchlings, mainly the latter)

Direct Threats

a. Incidental Catch in Fisheries

Fishing is a major occupation along the coast of India. A majority of fishers along the coast are artisanal fishers who use traditional methods or a combination of traditional and modern methods. However, since the 1970s, there has been a dramatic increase in mechanised fishing with thousands of motorised boats and trawlers operating in each state. The interface between marine fisheries and marine turtles has been a major concern not just for the wellbeing of sea turtle populations all over the world, but also for local and international commerce, artisanal fisheries, by-catch reduction policy, marine fishing ground health and the development of ecofriendly fishing gear.

Olive Ridleys Entangled in a Fishing Net

Several thousand turtles are killed in fisheries along the coasts of Orissa, Andhra Pradesh, Tamil Nadu and the Andaman and Nicobar Islands.

How Fisheries affect Turtles

Sea turtle mortality occurs primarily in gill nets and in trawl nets by drowning. Although sea turtles are capable of staying submerged for half an hour or more, the stress of being trapped for many hours in gill and trawl nets usually results in drowning. Some turtles trapped in trawl nets are not dead, but comatose, and if they are thrown back into the water immediately, they are likely to die. On the other hand, if they are kept on board the ship, they may recover. Turtle Excluder Devices (TEDs) can substantially reduce the mortality of turtles in trawl nets, without reducing fish catch.

In Orissa, more than a hundred thousand dead turtles have been counted since the 1990s, and ten to fifteen thousand dead turtles are washed ashore each year, predominantly due to trawlers. Researchers have documented that ten to twenty turtles can be trapped and killed during a single trawl. In 2002, researchers and conservationists also documented gill nets with 100–200 dead turtles trapped in each net. The real mortality rate may be even higher, since not all the turtles killed are washed ashore.

In addition, the operation of trawlers in offshore waters during the breeding season may disrupt breeding congregations and prevent the onset of mass nesting. Along the shore, many fishers operate zero mesh nets to catch shrimp seeds. These can interfere with the nesting of turtles and prevent hatchlings from reaching the sea.

Mitigation Measures

No Fishing Zones

Many maritime states already have laws to prevent mechanised fishing in certain zones, including all states on the east coast of India. Marine turtles are protected in Orissa by the Orissa Marine Fisheries Act (1982) and Rules (1983) which prohibit all mechanised fishing within 5 km of the coast. Fishing is also prohibited in the Gahirmatha Marine Sanctuary within 20 km of the Gahirmatha coast (~ 35 km). Currently, fishing is also restricted in the offshore waters of Devi river mouth in Rushikulya during the breeding season.

The Andhra Pradesh Marine Fishing (Regulation) Rules (1995) states that 15 m mechanised vessels may not operate within 8 km of the coast and vessels above 15 m in length may not operate within 25 km of the coast. Similarly, mechanised fishing is prohibited within 5 km of the coast in Tamil Nadu, and in the Gulf of Mannar Marine National Park. Similar laws exist for mechanised vessels in the state of West Bengal. However, implementation of these laws remains poor, and as a consequence, artisanal fishers, fish stocks and sea turtles all suffer.

Seasonal Fishing Ban

Many maritime states including Kerala, Tamil Nadu and Andhra Pradesh have seasonal fishing bans. These are currently during the south-west monsoon (May and June) to allow the replenishment of fish stocks. These bans could be extended for particular areas to the turtle breeding and nesting season as well.

Turtle Excluder Devices (TEDs)

A TED is a frame consisting of a grid of bars installed before the cod end of the trawl net at an angle leading upward or downward to an escape slit. Small animals such as shrimp slip through the bars and are retained in the cod end, while large animals, such as turtles, large fish and large elasmobranches are stopped by the grid bars and can escape through the opening. Many different kinds of soft and hard TEDs are available.

In India, the Central Institute of Fisheries Technology, Kochi, has developed an indigenous TED, called CIFT TED. The production of this TED is currently being funded by the Marine Products Export Development Authority. CIFT TEDs have been distributed in West Bengal, Orissa and Andhra Pradesh, but are not being used at any of the sites.

The use of TEDs has been made mandatory for trawlers in Orissa, and other states are following suit. The use of TEDs will be helpful, since they will prevent bycatch of many marine organisms, not just sea turtles. However, sea turtles and other fauna are also killed by gill nets.

Furthermore, the livelihoods of poor artisanal fishers must also be protected. In India, as in many parts of the world, the implementation of TEDs has been poor, as enforcement is difficult and the politics of fishing is complex. Hence a combination of TEDs and community supported seasonal and area wise fishing bans will probably give the best results. Illustration showing the working of a Turtle Excluder Device (TED).

b. Consumption of Adults

Along the Indian coastline, the take of adults and juveniles for consumption is not as common as in some other regions of the world. However, the practice does exist in many coastal villages.

Although such take and trade of turtle parts is in contravention of national (and international) legislation, often it is the ignorance of the law or the lack of support to, and understanding of, conservation initiatives that perpetuates the practice. It has been suggested that the involvement of the local community in conservation efforts by making them aware of the importance and benefits of such efforts will reduce their dependence on adult sea turtles as sources of food or income. It is important that the initiation of such programmes takes into consideration the socio-cultural dynamic of the area to ensure that they do not negatively affect the socio-economic structure of the community. Inclusion of the resource user community's interests is likely to increase the chances of their supporting the programme.

c. Egg Predation

Predation of sea turtle eggs is a common threat in nesting beaches around the world. Eggs are predated by natural and subsidised predators, and in many instances by humans.

Feral dogs and crows are common predators of sea turtle eggs.

Indirect Threats

a. Threats in Offshore Waters

With the onset of the breeding season, sea turtles leave their foraging areas and arrive in the coastal waters of breeding grounds. Species like olive ridleys congregate in large numbers in the coastal waters off nesting beaches and spend nearly six months in a year in this habitat.

Green and hawksbill turtles use shallow water lagoons as their developmental habitat during the juvenile and sub-adult stages of their life.These nearshore waters are often subject to heavy use by humans in terms of marine fishing, aquaculture, coastal tourism and other recreational activities. These coastal zones are alsosusceptible to indirect pollution from industrial or agricultural run-off. These activities impact sea turtles during various crucial stages of their life cycle.

Pollution of Offshore Waters

Coastal waters and beaches are under constant threat from pollution originating from landward activities as well as from the sea. Such pollutants have often proved to be the primary cause of death of marine plants and animals. Impacts of pollution on sea turtles have been less severe, although accidental ingestion of plastics has been documented to indirectly cause death due to poisoning or starvation because of the inability to swallow food due to the blockage of the food passage by these materials. Further, changes in water temperature and quality results in changes in their offshore breeding congregation locations. Changes in salinity profile and levels of organic and inorganic pollutants in the vicinity of mass nesting sites will impact adults and hatchlings.

The following pollutants are known to impact sea turtles directly or indirectly through their food chains or altering their preferred offshore/onshore habitats:

1. Discharge of sewage, nutrients and fertilisers: In sheltered lagoons, organic sewage, nutrients andfertilisers tend to accumulate and hasten the process of algal bloom and seaweed growth, changing the food chain composition.
2. Heated water from power stations and industrial plants: In lagoons and offshore waters, release of heated water from power plants and other industrial installations changes the water temperatures significantly altering the distribution pattern of microorganisms. This in turn impacts the distribution of organisms at various trophic levels and also that of sea turtles.
3. Petroleum hydrocarbons and crude oil from oil spills: Oil spills cause severe damage to marine life worldwide due to oil pollution along the

coasts. Sea turtles are exposed to the harmful effects of oil pollution in many ways. Floating crude oil in the immediate offshore waters gets coated on the eyelids, nostrils and mouths of sea turtles, often leading to death. This has been recorded along the west coast of India. On nesting beaches, oil deposits could interfere with proper aeration of the turtle nests and impair normal development of the embryos. The crude encrusted hard substrate also does not allow the embryos to successfully emerge from the nest. Onshore oil deposits are high along the west coast of India, particularly along the Gulf of Kutch where the presence of many oil refineries have resulted in chronic crude oil spills into the sea.

4. Oil deposits on nesting beaches from spills impact both nesting turtles and nests
5. Chemicals and synthetic paints, pesticides and herbicides, and heavy metals: These may selectively destroy or damage phytoplankton and zooplankton of reef and lagoon communities as well as planktonic larvae. Such accumulation has severe physiological effects on filter feeding animals and reef fish and may be accumulated in animals like the sea turtles. Pesticides and other industrial chemicals have been identified as "environmental estrogens". These can impact the reproductive system of animals including turtles if the beach sand becomes contaminated with these types of chemicals.
6. Radioactive waste: In recent years, research has shown marine turtles exhibiting carcinogenic growths and deformities attributed to radioactive residues. There is also a suspicion that radioactive waste may result in long term and largely unpredictable effects on the genetic nature of the biological communities in the sea, including sea turtles.
7. Debris: Marine debris is prevalent in nearshore habitats, and there are numerous reports of the occurrence of marine debris in the digestive tracts of hatchling and adult sea turtles. The list of materials found in the digestive track of sea turtles is extraordinary. Plastic bags, sheets, beads, pellets, lines, rope, strapping, pieces from bottles and hard pieces of unknown origin are commonly ingested by sea turtles. If sufficient material is swallowed it can cause complete stoppage of the gut and result in death. Leatherbacks feed principally on jellyfish and are known to swallow plastic bags by mistake, often leading to their death.

Mitigation Measures

Options to minimise pollution include:

- ☆ Monitoring of levels of pollution both onshore and offshore, particularly in sea turtle congregation areas
- ☆ Prohibition of discharge of crude oil, pesticides, heavy waters, heavy metals and other poisonous effluents to estuaries and coastal areas and near turtle nesting beaches

- ☆ Prohibition of discarding of fishing lines, nets, plastic bags and other trash into the water or on the beach which results in ghost fishing and incidental mortality of sea turtles
- ☆ Introduction of organisedclean up of beaches and nearshore waters by local forest and fisheries departments and other governmental agencies in collaboration with non-governmental organisations, coastal communities, and school and college students. This should particularly be taken up prior to the breeding season of sea turtles in an attempt to clean up the habitat as well as to educate people about marine turtles

b. Threats to Nesting Beaches

i. Sand Mining

The coastal and nearshore marine environment, sea sand in particular, is a source of a variety of minerals of geological and biological origin that have been extracted and utilised by man for centuries. In general, marine sand used for building and construction are utilised locally, whereas those products that have a world market, such as jewelry and industrial metals, may be largely exported in one form or other. The mining or extraction of these minerals, however, tends to be unplanned and unmanaged, causing severe and long-lasting detrimental impacts to the environment.

Beaches are dynamic landforms and are constantly subject to erosion and/or accretion. The sand on the beach is subject to storms, waves, and buffeting from the force of waves. This in turn results in the movement of sand from one part of the beach by erosion, and accretion of the same sand in another part. The condition of a beach is a reflection of the local balanced or unbalanced gain due to deposition or loss due to erosion. While natural beachfront and sea erosion does occur, anthropogenic alteration of the beach has significantly contributed to beachfront erosion.

Therefore, beach conservation should be based on the premise that any removal of sand can have adverse impacts by disrupting the natural cycle.

Impact of Sand Mining on Sea Turtle Nesting Beaches

Coastal sand mining can change the entire beach geomorphology. Restoration of the beach often takes years, resulting in loss of available habitat for marine flora and fauna. The immediate deleterious impact of beach sand mining on sea turtles is the uncovering and destruction of nests. Raking can also leave ruts and ridges that disrupt hatchlings' sea finding behaviour. Beach nourishment is one solution to counter the impacts of sand mining, but it can negatively impact sea turtles if the sand is too compacted for nesting. If the sand imported is drastically different from native beach sediments, it may affect nest-site selection, digging behaviour, incubation temperature, gas exchange and moisture content of nests, all of which can ultimately impact the reproductive fitness of sea turtles.

In India, there is severe damage to the nesting beaches of olive ridley turtles along the coast of Orissa (sand mining for rare earth metals), Andhra Pradesh (sand mining for minerals), and Kerala (sand mining for building constructions). In the

Andaman and Nicobar Islands, in the absence of riverine sand, large scale sand mining from narrow fringes of sandy coastline impacts nesting beaches of olive ridley, green, hawksbill and leatherback turtles.

Guidelines for Coastal Sand Mining

The Government of India, in 1991, issued the Coastal Regulation Zone (CRZ) Notification under the Envrionment Protection Act (EPA) of 1986 to regulate development activites such as sand mining, lanscaping and construction along coastal stretches of mainland India. The notification has since been amended numerous times and the latest amendement was was officially passed in 2011. The Andaman and Nicobar and Lakshadweep Islands are governed by a recent notification (also issued under the EPA) titled Island Protection Zone (IPZ) Notification.

Coastal areas which include the nesting and breeding grounds of endemic and endangered species are categorised under both the CRZ and IPZ notifications as requiring a high degree of protection, and as such, no mining of sand is allowed at these sites. Both notifications contain a list of activities that are permitted and prohibitted along designated coastal stretches and contain provisions for coastal planning through the preparation of Coastal Zone Plans. Under the CRZ 2011 and IPZ 2011, mining of sand, rocks and other sub-strata materials is prohibited, except for rare minerals not available outside the CRZ area and for exploration of oil and natural gas.

Before taking up any coastal sand mining activity, the concerned agencies must ensure that the targeted areas are not important habitats of any marine fauna. Sensitive coastal stretches need to be identified, properly marked with site boards and labels, and removal of sand from such zones or primary dunes should be completely prohibited. Even where sand mining may be allowed, the actual mining must be preceded by proper impact assessment studies conducted by the development agencies in collaboration with the coastal resource management authorities including environment, forest and wildlife agencies. If sand extraction from the landward side of the berm is permitted, it should be carefully conducted avoiding the main nesting season of sea turtles and other rare and endangered fauna, and hard engineering exercises must be stopped at night.

Sand mining activities must ensure adequate control to prevent sedimentation of watercourses from spoil deposits and other disturbance on the land surface. Sand fill and other restoration processes need to be monitored to confirm whether such measures actually rehabilitate sand dunes, and whether the new fill is compatible with the needs of nesting sea turtles and other coastal fauna and flora.

ii. Beach Armouring

The deliberate 'armouring' of the coast with the sole purpose of protection of upland structures is rapidly degrading sea turtle nesting habitats in many parts of the country. Coastal armouring takes many different forms such as bulkheads and seawalls, revetments, sandbags and geotextile tubes, soil retaining walls and dune reconstruction. Beach erosion has been a problem especially along the west coast of India. To mitigate erosion, beaches have been armoured with concrete tetrapods and sand bags along the Kerala and Karnataka coasts, Lakshadweep and Andaman and Nicobar Islands. Wherever coastal development has been planned and the area is subjected to some degree of erosion, concrete armouring has been carried out without conducting any impact assessment studies on marine flora and fauna (see Namboothri *et al.*, 2008a, b; Rodriguez *et al.*, 2008).

Beach Armouring in the Lakshadweep Islands

Possible impact on sea turtles

There are four broad consequences to the beach/dune system that can result from coastal armouring.

1. Coastal armouring structures cause reflection in wave energy, which can increase erosion seaward of these structures.
2. The intensity of long shore currents can be increased, moving sand away from the site more rapidly and in greater quantities.
3. The natural exchange of sand between the dune and the beach is prevented; the wave energy is concentrated at the end of armoring structures, which can exacerbate erosion at an adjacent, unarmoured beach.
4. These structures physically block female turtles from reaching suitable nesting sites, or the presence of these structures may be aversive to nesting females. These structures may also disrupt sea finding of hatchlings once they emerge.

Mitigation Measures

Guidelines for ecofriendly armouring:

- ✰ All important sea turtle nesting beaches of moderate intensity must be identified and should be free from beach armouring as sea turtles normally prefer gentle sloping seaward sand dunes rather than beaches which are subject to regular erosion.
- ✰ Beach changes must be measured and monitored on a regular basis; this information should be used by planning agencies and others to reduce the problems caused by coastal erosion and to effectively manage coastal development.
- ✰ Construction of temporary sand trapping fences: A typical sand fence consists of vertical slats joined with wire/rope or supported with sand posts. Such structures have proved to be more effective in controlling beach erosion and are relatively sea turtle friendly. Caution must be exercised when selecting the placement of these fences so as to avoid impacting nesting females. In general, sections of fence should be less than 3 m in length with a 45° angle relative to the high tide line. The sections of fence should all be parallel to each other.
- ✰ Sand trapping fences may be constructed in destroyed, eroded beaches during the nonnesting season and once the sand is trapped sufficiently

and a dune is created, it should be planted with appropriate vegetation. This will further reduce erosion and remove the need for armouring. Depending on availability, the fence can be constructed with discarded coconut branches, dry wood and logs, and other materials. The fence must be stabilised with vegetation.

- ☆ Coastal vegetation promotes the large scale trapping of sand. The stems of beach grasses reduce the wind velocity near the surface resulting in deposition of sand. The plant roots also serve to bind and consolidate the sand. Therefore, re-vegetation is a suitable option for eroded beaches where armoring is planned. However, this should be area specific and depend upon the characteristics of a particular site. For example, olive ridleys prefer wide sandy beaches, and no vegetation should be planted up to a distance of 50-100 m from the high tide line. On the other hand, green turtles prefer to nest in coastal undergrowth vegetation rather than on open sandy beaches.

Highways End Marine Drives

In the process of construction of highways, coastal geomorphology is considerably altered.

Engineering actions are often geared towards the highways rather than towards the natural coastal geomorphology and ecological processes. Deltaic drainages, coastal vegetation and the shore sand dunes often get impacted irreversibly. Apart from reducing natural coastal habitat, a highway also brings in new impact factors not present in the area before the construction of the highway. In the context of marine turtles, coastal highways on sand dunes have invaded sea turtle nesting sites, and highway illumination and vehicular traffic illumination have impacted marine turtles directly. With increasing access to the coast, human settlements crop up on the coast and along with humans, accompanying livestock and pets (dogs) cause considerable damage to the coastal ecosystem. Predation on adults and hatchlings gets enhanced significantly as the presence of subsidised predators increases.

Impacts of Roads on Sea Turtles

Roads that are constructed on the dune and/or very close to the high tide line of the sea, directly impact sea turtles by:

- ☆ Reducing the space available for them to nest.
- ☆ Disturbing their egg laying activities (due vehicular traffic during the night).
- ☆ Increasing human and other biotic disturbance on the nesting beach.
- ☆ Removing beach sand for road construction, thereby eliminating significant amounts of nesting habitat.
- ☆ Altering the geomorphology of the sand dunes and reducing the natural nesting beach over a period of time.

Mitigation Measures

The impacts of coastal highways can be reduced by:

- ✰ Locating the road sufficiently far away from the coastal sand dune habitat, preferably on the landward slope of the sand dunes rather than the seaward slope.
- ✰ Plantation of indigenous species on the seaward side of the highway to reduce lights on the beach from vehicular traffic and marine drives and also to safeguard the highway from sand deposition.
- ✰ Closure of vehicular traffic at night, at least during nesting season at prime sea turtle nesting areas.
- ✰ Creation and restoration of degraded sand dune habitats caused by highway construction.

iii. Exotic Plantations

While the beach itself is devoid of any vegetation due to the constant action of waves, the sand dune zone above the high tide line has various types of natural vegetation despite being subject to wind action. Sand dune grass and other spiny vegetation trap blowing sand and help stabilise, maintain and elevate dune structures. There are also other salt tolerant creepers that help bind the sand.

Most sea turtles prefer open sandy beaches for nesting. Coastal sand dunes with natural vegetation such as Ipomea pes-caprae and Spinifexlittoreus are ideal nesting sites for sea turtles in India. The natural dune vegetation and sand dune structures are closely interrelated and have coevolved to support floral and faunal communities obligate to the coastal sand dune habitat.

However, in recent years, as a measure of control of beach erosion, creation of vegetation shelterbelts against cyclonic storms and afforestation of the coastal zone, large-scale plantations of alien and exotic plant species have been taken up, without any impact assessment studies.

One of the exotics that has been planted along most of the Indian coast is the Australian screwpine *Casuarina equisitefoli*. Casuarina plantations were also widely established on the east coast of India (especially Andhra Pradesh and Tamil Nadu) following the December 2004 tsunami. However, studies have shown that these plantations, unlike natural mangrove forests, can hardly withstand cyclones and storm surges

Impact of Coastal Plantations on Sea Turtles

Exotic coastal plantations have proved to be detrimental to the nesting of sea turtles in more ways than one.

- ✰ Plantation of exotic beach vegetation drastically alters the beach profile and may often be a deterrent for sea turtle nesting.
- ✰ Since all coastal plantations are taken up on either side of the sand dune berm, there is significant loss of sea turtle nesting habitat between the high tide line and the berm on the seaward side.

- Alien vegetation such as Casuarina with its superficial root growth and thick litter fall renders the beach unsuitable for turtles to nest.
- Dense Casuarina and other plantations cause excessive shading of the nesting beach. Nests laid in shaded areas are subject to lower incubation temperature, which alters the natural sex ratio of turtle hatchlings, producing more males.
- Plantation of exotic vegetation on the beach also affects the natural beach formation process.
- Dense vegetation on the coastal sand dunes provides shelter for both natural (jackals, hyenas, monitor lizards, wild pigs, etc.) andsubsidised predators (dogs, pigs, etc.) to breed and add additional predation pressure on nesting females, eggs and hatchlings.

In India, some of the important mass nesting beaches of olive ridleys along the coast of Orissa have been drastically altered with dense Casuarina plantations. Some nesting beaches of Gahirmatha and Devi river mouth that were used by olive ridleys in extremely large numbers in the early 1970s and 1980s have been completely abandoned by sea turtles after dense strands of Casuarina were planted.

Guidelines for Ecofriendly Coastal Vegetation

The biological and ecological significance of the beach is often overlooked while undertaking developmental activities or afforestation programmes. The following guidelines are therefore suggested for ecofriendly revegetation of the coastal sand dunes.

- As a policy, no exotic species should be planted in CRZ areas so as to prevent unknown/unforeseen ecological impacts.
- Environmental and ecological impact assessment studies must be conducted before taking up any afforestation programmes along the coast, with provisions set up to evaluate any post-project impacts on sea turtles.
- Plantation of exotic vegetation should be clearly avoided on beaches that are known to be sea turtle nesting grounds.
- Apart from the high priority sea turtle nesting beaches where no plantation should be taken up, in sporadic sea turtle nesting areas, plantations should be carried out beyond 200 m from high tide line (HTL) and on the landward side slope of the berm, thereby setting aside enough space for sea turtles to nest.
- In historically known sea turtle nesting sites, alien and exotic plantations should be gradually removed to restore the nesting beaches to their former state.

iv. Artificial Illumination

The emergence of adult females to successfully nest on a particular beach depends on many factors, including the natural suitability of the nesting site and presence or absence of abiotic and biotic disturbances. With increasing use of the coast for developmental activities, nesting sites are often abandoned by sea turtles due to human induced activities, including artificial coastal illumination.

Hatchlings use visual cues to find the sea when they emerge. Both nesting adults and hatchlings therefore, are susceptible to changes brought on by the use of artificial lighting on the beach.

Impact of Coastal Illumination on Sea Turtles

Impact on Adults

Artificial illumination on nesting beaches impacts adult sea turtles by disrupting nest site selection, abandonment of nesting behaviour, disruption of sea finding ability and disorientation following unsuccessful nesting. The most clearly documented effect of artificial lighting on sea turtle nesting beaches has been the non-emergence of adult females to the nesting beach.

Dramatic reductions in nesting attempts by sea turtles at brightly lit nesting beaches have been documented for all species of sea turtles. Along the Indian coastline, conservationists have attempted to reduce artificial illumination near Gahirmatha and Rushikulya in Orissa during the nesting season, to prevent disruption of nesting by olive ridley turtles.

Sea turtles are particularly sensitive to disturbance during the initial phase of nesting. Generally, female turtles will abandon egg laying and return to the sea if disturbed by lights and other activities. After nesting, sea turtles may move towards the brightly illuminated landward side of the seashore rather than going back to the sea. Such bright illumination along the seashore often comes from highways, beach resorts, industries, coastal village street lights and townships, ports and jetties and in recent years, coastal aquaculture farms.

Impact on Hatchlings

Sea turtle hatchlings orient themselves towards the sea as soon as they emerge from the nest.

Under natural conditions, the hatchlings recognise the direction of the ocean almost exclusively by visual stimuli, detecting the brightness of the open seaward horizon, due to the reflection of stars and moonlight on water. Sand dunes and vegetation along the nesting beach also help create a darker horizon on the landward side. However, on beaches where artificial lighting is clearly visible, the hatchlings' journey to the sea is disrupted. Hatchling sea turtles emerging from nests at night are strongly attracted to visible light sources along the beach. Consequently, hatchlings move toward the source of artificial illumination and away from the ocean. They thus fail to find their way to the sea, and succumb to predators or exhaustion or dehydrate in the morning sun. Their orientation towards the artificial lights is so strong that even a small torch light left on the beach can attract hundreds of hatchlings towards it.

Some of the major nesting sites of olive ridleys in India like Rushikulya in Orissa are now subject to heavy artificial illumination. This has resulted in heavy mortality of turtle hatchlings.

At the Rushikulya mass nesting beach in some years, hundreds of thousands of hatchlings have strayed into the fields and vegetation behind the beach. These are collected in the morning by conservation volunteers, state Forest Department staff and local communities, for release into the sea. With the present trend in coastal development, artificial illumination is going to pose a serious impediment for sea turtles along the entire mainland coast, and even if they nest successfully, the hatchlings may stray onto land and die.

Guidelines for Coastal Illumination

Any light that is visible to the human eye at turtle level on a beach affects the nesting turtles and/or hatchlings. The most direct and complete way to resolve the problem of lights near sea turtle nesting beaches is to put off all artificial lights visible on the nesting beaches during the breeding season. Unfortunately eliminating all the beach front lighting is not always possible.

Under such circumstances, following a few simple measures can help mitigate impacts.

- ☆ Artificial illumination along important sea turtle nesting beaches during the nesting season must be turned off (depending on the species, the peak nesting season at any particular location usually does not exceed three months of the year).
- ☆ The number of lights near sporadic and secondary nesting beaches must be reduced to the minimum necessary and switched off during peak nesting nights.
- ☆ Illumination reaching the nesting beach can be reduced by lowering, shielding, and redirecting light sources onto immediate land rather than towards the sea. Even the glow on the horizon can affect sea turtles. Low mounted lights are better than lights that shine upwards from a high pole.
- ☆ An easy means to reduce the influence of light on turtle beaches is to screen them on the seaward side. This can be done by applying dark tinting to windows visible on the beach and by drawing curtains after dark to shield lights.
- ☆ Care in placement and orientation of light fittings on coastal buildings and infrastructure will considerably reduce the impact of direct and scattered lighting on turtle nesting beaches. Orienting light towards land or by directing lights downwards and by using lamp shades is an easy and effective solution.
- ☆ Studies have shown that fluorescent, mercury vapour, high pressure sodium vapor, metal halide and white incandescent lighting disorient sea turtles the most. Further studies need to test which lights affect sea turtles the least (and this may be species specific). For loggerheads, low pressure sodium vapor lights (not to be confused with high pressure sodiumvapour lights) that emit a pure yellow light seem to work best. Similar studies need to be carried out for olive ridleys which nest along much of the Indian coast and are most affected by artificial lighting.

- ☆ Security lighting can be placed on motion sensitive switches that keep lighting off when it is not needed. Lights that come on only when approached can be quite effective for security purposes.
- ☆ Vegetation can suitably "block" artificial lighting from reaching the beach. This might not be applicable in all situations, but may work in specific cases (see Karnad *et al.*, 2009). There is a definite need for a Coastal Illumination Act to define appropriate beachfront lighting, to set standards, apply regulations, and deal with offenders. Compared to other kinds of coastal development, light pollution is probably one of the problems that can be solved with relative ease, if the government and local residents are committed to conserving sea turtles.
- ☆ Total prohibition on the release of chemicals and effluents from aquaculture farms and hatcheries into marine and coastal riverine systems.

v. Tourism

The beachfront and the immediate adjacent land, lagoons and offshore waters are preferred locations for coastal tourism. One of the major impacts of such development is the drastic alteration of the natural landscape often completely stopping, or interfering with, ecological processes. An ecofriendly coastal tourism strategy and action plan should seek to optimise developmental benefits while preserving the natural environment and the socio-cultural scenario upon which the sector depends.

Impact of Coastal Tourism on Sea Turtles

Tourism infrastructure: Physical alteration and loss of nesting beaches comes about as a result of erection of structures such as hotels, apartments, restaurants, mobile shops, etc., on the beachfront. They also disorient adults and hatchlings due to artificial illumination. Shadows from buildings close to the high tide line can alter sand temperatures and thus affect the sex ratios of hatchlings.

Removal of vegetation, sand, corals etc.: Removal of vegetation, sand and other materials for recreational tourism results in erosion of sand from nesting beaches and makes the beach unsuitable for sea turtles to nest. Green and hawksbill turtles prefer to nest in the shade of coastal vegetation and if vegetation is removed, they may not use such beaches for nesting. Speedboat movement and anchoring: Small propeller driven recreational and other commercial boats used by inexperienced boat handlers often cause considerable physical damage to shallow marine habitat including reefs, particularly at low tide. A large number of marine turtles succumb to propeller related injuries. Anchoring of boats in shallow water areas, particularly in corals and seagrass beds, damages these habitats. Plough anchors are known to be particularly destructive. Oil spillage and chronic leakages from tourist and other boats are also a major source of pollution in lagoons and nearshore waters.

Beach chairs, sun beds, umbrellas, etc.: Beach chairs, umbrellas etc. on the beach stop female turtles from reaching suitable nesting locations. Installations of such artifacts can also damage the nest on the beach and/or interfere with incubation temperatures due to increased shade.

Mechanical cleaning: Mechanical cleaning of the beach contributes to the compaction of sand and destroys turtle nests. Further, mechanical cleaning changes the beach slope and configuration thereby deterring adult female turtles from using the area for nesting.

Joy rides on the beach using animals: Use of heavy animals for joy rides on the beach can destroy turtle nests by trampling, and continuous movement of these animals on the beach also results in compaction of beach sand which makes the beach unsuitable for nesting and emergence of hatchlings.

Littering of beach by tourists: Trash and debris on the shoreline not only threaten the health and safety of beach users but also entangle marine animals, including sea turtles. Debris such as plastic sheets, scraps of nylon nets and ropes often cause entanglement of nesting turtles and hatchlings.

Snorkeling and scuba diving: Snorkeling and scuba diving by tourists in turtle congregation areas disrupts the foraging and breeding activities of turtles. Since such activities are often near coral reefs, they also cause unintentional damage to corals and other reef biota resulting in significant loss of sea turtle habitat.

Lagoon/coral bed fishing: Recreational fishing in lagoons and coral beds, harpooning, use of explosives, collection of ornamental fish, live corals, shells, seagrass and seaweeds not only hampers major activities of sea turtles but also reduces forage availability and causes considerable damage to these habitats.

Guidelines for (Turtle) Eco-friendly Coastal Tourism

- ✰ Seasonal closure of coastal tourism activities in nesting beaches and lagoons should be adopted.
- ✰ Tourism infrastructure development should only be permitted beyond 200 m from the high tide line or on the landward slope of the sand dune berm rather than the seaward slope.
- ✰ All illumination must be made sea turtle friendly. Green ratings, and punish and award systems can be adopted to encourage turtle friendly infrastructure.
- ✰ A total ban on joy rides on nesting beaches using animals or heavy vehicles should be enforced.
- ✰ Beach leveling and removal of natural vegetation should only be permitted in consultation with conservationists or coastal zone management authorities.
- ✰ Foot traffic on sea turtle nesting beaches should be controlled during the daytime to avoid compaction of sand, and at night to avoid disturbing nesting turtles, in those areas where tourists are not taken under well organised sea turtle watch programmes.
- ✰ The tourism department, beach resorts and other beneficiaries of beach tourism can involve local communities and schools to adopt a particular sea turtle nesting beach and demonstrate turtle friendly practices.

Installation of Permanent or Temporary Structures Impact Turtle Nesting Veaches

The following provisions must be made on sea turtle nesting beaches where tourism is also an important focus: provision of litterbins, installation of warning notices and boards, demarcation of turtle sensitive areas, and display of guidelines for lagoon and offshore habitat users. Innovative methods can be used to promote sea turtles as a coastal tourism resource, for example, by adopting turtle friendly tourism such as turtle watch in lagoons, nesting turtle watch, turtle egg collection, hatchery operation and release of hatchlings and voluntary nesting turtle counts. Such activities are not only educative but also generate a great deal of interest and support for sea turtle conservation.

'Operation Kachhapa'; Conservation of the Olive Ridley Sea Turtle

The State of Orissa on India's eastern coast witnesses a spectacular natural phenomenon – the arribada or mass nesting of hundreds of thousands of Olive Ridley sea turtles. The turtles first congregate off shore to mate. Then, on a few moonlit nights during December to March, thousands of females land on the beach to lay eggs and then return to the sea.

Unfortunately, on the Orissa coast, about 16,000 fishing boats operate during the winter season. Some 8,000 of these boats are large mechanised fishing trawlers and gill-netters that flout laws prohibiting mechanised fishing within five kilometres of the shoreline. Turtles are injured by the boats' propellers and slaughtered in the fishing nets – a staggering total of 100,000 dead turtles have been washed ashore in the past decade, almost all of them breeding adults.

"It is a tragedy with irreversible implications. No species can survive such high mortality of its breeding populations" says Belinda Wright.

WPSI launched Operation Kachhapa in September 1998 to try to stem the slaughter. objective is to reduce turtle mortality and try to safeguard the future of the species by concentrating on three main activities:

1. To improve patrolling of non-fishing zones and the protection of nesting sites.
2. To support legal action on turtle conservation and fishing law violations.
3. To build public support and awareness of sea turtle conservation issues.

In collaboration with the Orissa State Forest Department and the Wildlife Society of Orissa and other local NGOs, programme activities have included:

1. Providing the State enforcement authorities with sea-going patrol boats, that have seized hundreds of trawling and gill-netting vessels;
2. Organising field camps to protect and monitor turtle movements;
3. Carrying out sea turtle mortality counts;

Providing legal support to:

1. Protect the sea turtles and their habitat
2. Restrict illegal mechanised fishing
3. Encourage trawlers to use mandatory turtle excluder devices (TEDs);
4. Organising multi-language awareness programmes;
5. Setting up a Turtle Interpretation Centre in Bhubaneshwar, Orissa's capital, to help spread awareness of the need for conservation activities;
6. Providing logistical support to the UNDP Turtle Project.

A special turtle song was composed in Oriya, Orissa's local language, so that traditional methods of communication such as song, dance and story telling could be used in the awareness programme. Two wandering minstrels have performed the song in over 160 villages.

Operation Kachhapa has had a significant impact on the number of sea turtle deaths and public awareness of the problem. But construction activity along the coast is relentless and sadly thousands of turtles still die every year from the the illegal fishing boats that ply off the coast of Orissa. The future of the Olive Ridley sea turtle is still a matter of great concern.

Why Healthy Oceans needs Sea Turtle

Sea turtles are part of two vital ecosystems, beaches and marine systems. If sea turtles become extinct, both the marine and beach ecosystems will weaken. And since humans use the ocean as an important source for food and use beaches for many kinds of activities, weakness in these ecosystems would have harmful effects on humans.

Though sea turtles have been living and thriving in the world's oceans for 150 million years, they are now in danger of extinction largely because of changes brought about by humans. If we alter the oceans and beaches enough to wipe out sea turtles, will those changes make it difficult for us to survive? And if we choose to do what's necessary to save sea turtles, might we save our own future?

Beaches and dune systems do not get very many nutrients during the year, so very little vegetation grows on the dunes and no vegetation grows on the beach itself. This is because sand does not hold nutrients very well. Sea turtles use beaches and the lower dunes to nest and lay their eggs. Sea turtles lay around 100 eggs in a nest and lay between 3 and 7 nests during the summer nesting season. Not every nest will hatch, not every egg in a nest will hatch, and not all of the hatchlings in a nest will make it out of the nest. All the unhatched nests, eggs and trapped hatchlings are very good sources of nutrients for the dune vegetation. Even the left-over egg shells from hatched eggs provide nutrients.

Dune plants use the nutrients from turtle eggs to grow and become stronger. As the dune vegetation grows stronger and healthier, the health of the entire beach/ dune ecosystem becomes better. Healthy vegetation and strong root systems hold the sand in the dunes and protect the beach from erosion. As the number of turtles declines, fewer eggs are laid in the beaches, providing less nutrients. If sea turtles went extinct, dune vegetation would lose a major source of nutrients and would not be healthy or strong enough to maintain the dunes, allowing beaches to wash away.

Sea turtles eat jellyfish, preventing the large "blooms" of jellyfish – including stinging jellyfish – that are increasingly wreaking havoc on fisheries, recreation and other maritime activities throughout the oceans.

Research has shown that sea turtles often act as keystone species. Sea grass beds grazed by green sea turtles are more productive than those that aren't. Hawksbill turtles eat sponges, preventing them from out-competing slow-growing corals. Both of these grazing activities maintain species diversity and the natural balance of fragile marine ecosystems. If sea turtles go extinct, it will cause declines in all the species whose survival depends on healthy seagrass beds and coral reefs. That means that many marine species that humans harvest would be lost.

Sea turtles, and many species that are affected by their presence or absence, are an important attraction for marine tourism, a major source of income for many countries.

References

Aggarwal, K.R., T.P. Velavan, D. Udaykumar, P.S. Hendre, K. Shanker, B.C. Choudhury and L. Singh. 2004. Development and characterization of novel microsatellite markers from the olive ridley sea turtle (Lepidochelysolivacea). *Molecular Ecology Notes* 4: 77–79.

Andrews, H.V. and K. Shanker. 2002. A significant population of leatherback turtles in the Indian Ocean.*Kachhapa* 6: 17.

Andrews, H.V., S. Krishnan and P. Biswas. 2006. Distribution and status of marine turtles in the Andaman and Nicobar Islands. In: Marine turtles of the Indian subcontinent (Eds. Shanker, K. and B.C. Choudhury). pp. 33–57. Hyderabad: Universities Press.

Benton, Michael J., Ph.D. The Reign of the Reptiles. New York: Crescent Books, 1990.

Bhaskar, S. 1993. The status and ecology of sea turtles in the Andaman and Nicobar Islands. Centre for Herpetology/Madras Crocodile Bank Trust. Publication: ST 1/93.

Bhupathy, S. and S. Saravanan. 2002. Status survey of sea turtles along the Tamil Nadu coast. A GOI-UNDP Sea Turtle Project Report. Salim Ali Centre for Ornithology and Natural History, Coimbatore, India.

Bjorndal, K. (Ed.). 1995. Biology and Conservation of sea turtles. Washington D.C.: Smithsonian Institution Press.

Bjorndal, K.A. 2003. Roles of loggerhead sea turtles in marine environments. In Bolten, A.B. and Witherington, B.E. (Eds.) Loggerhead Sea Turtles. Smithsonian Institution Biology 16(13): R492-R493.

Bjorndal, K.A., Jackson, J.B.C. and Preen, A. R. 1996. Infaunal mining: A novel foraging method of loggerhead turtles. *Journal of Herpetology* 30(1): 94-96.

Blassingame, Wyatt. Wonders of the Turtle World. New York: Dodd, Mead and Co., 1976.

Bluvias, J.E. and Eckert, K.L. (Adviser). 2008. Marine turtle trauma response procedures: A husbandry manual. Master's thesis submitted to Nicholas School of the Environment and Earth Sciences of Duke University, USA.

Bouchard, S.S. and Bjorndal, K.A. 2000. Sea turtles as biological transporters of nutrients and energy from marine to terrestrial ecosystems. *Ecology* 81: 2305-2313.

Bustard, Robert. Sea Turtles: Their Natural History and Conservation. New York: Taplinger Publishing Co., 1972.

Carr, A.F. 1967. So excellent a fishe: A natural history of sea turtles. New York: Natural History Press.

Carr, Archie. Handbook of Turtles. London: Cornell University Press, 1952.

Carr, Archie. So Excellent a Fish. Garden City, New York: The Natural History Press, 1967.

Chaudhari, S.A. K.V. Devi Prasad and K. Shanker. 2009. Impact of Casuarina plantations on olive ridley turtle nesting in Tamil Nadu. Ashoka Trust for Research in Ecology and the Environment, Bangalore and Madras Crocodile Bank Trust, Mamallapuram, India.pp. 44.

Choudhury, B.C. 2008. Ecology of marine turtles in the Lakshadweep Islands, India, with a focus on Hawksbill turtles (Eretomochelys embricata). Wildlife Institute of India, Dehradun.

Clark, J.R. 1996. Coastal zone management handbook.694 pp. New York: Lewis Publishers.

Dixon, Dougal, *et al.* Macmillan Illustrated Encyclopedia of Dinosaurs and Prehistoric Animals. A Visual Who's Who of Prehistoric Life. New York: Macmillan Publishing Co., 1988.

Eckert, K.L., K.A. Bjorndal, F.A. Alberto Abreu-Grobois and M. Donnelly (Eds.). 1999. Research and management techniques for the conservation of sea turtles. IUCN/SSC Marine Turtle Specialist Group Special Publication No. 4.

Ernst, Carl H. and Roger W. Barbour.Turtles of the World. Washington, D.C.: Smithsonian Institution Press, 1989.

Fergusson, I. K., Compagno, L. J. V., Marks, M. A. 2000 Predation by white sharks Carcharodon carcharias (Chondrichthyes: Lamnidae) upon chelonians, with new Fishes 58(4): 447-453.

Gibbons, M.J. and Richardson, A.J. 2009. Patterns of jellyfish abundance in the North Atlantic. *Hydrobiologia* 616:51-65.

Gopi, G.V., B.C. Choudhury and B. Pandav. 2002. A quantative analysis of incidental sea turtle captures and mortalities during commercial shrimp trawling in the coastal waters of Orissa. Wildlife Institute of India, Dehradun.

Gulko, D.A. and Eckert, K.L. 2004. Gyuris, E. 1994. The rate of predation by fishes on hatchlings of the green turtle (*Chelonia mydas*). Coral Reefs 13:137-144.

Gulko, D.A. and Eckert, K.L. 2004. Sea Turtles: An Ecological Guide. Mutual Publishing, Honolulu, HI. 128 pp. Ser. 350: 153-174.

Gulko, D.A. and Eckert, K.L. 2004.The Biology of Sea Turtles. CRC Press, Boca Raton. pp. 137-163.

Hannan, L.B., Roth, J.D., Ehrhart, L.M., and Weishampel, J.F. 2007. Dune vegetation fertilization by nesting sea turtles. *Ecology* 88(4): 1053-1.

Heithaus, M.R., Frid, A., Wirsing, A.J., Dill, L.M., Fourquerean, J.W., Burkholder, D., Thompson, J., Bejder, L. 2007. State-dependent risk-taking by green sea turtles mediates top-down effects of tiger shark intimidation in a marine ecosystem. *Journal of Animal Ecology* 76(5):837-844.

Hirth, H.F. Synopsis of Biological Data on the Green Turtle. Rome: Food and Agriculture Organization of the United Nations, 1971.

Houghton, J.D.R., Doyle, T.K., Wilson, M.W., Davenport, J., and Hays, G.C. 2006. Jellyfish aggregations and leatherback turtle foraging patterns in a temperate coastal environment. *Ecology* 87(8): 1967-1972.

Leon, Y.M. and Bjorndal, K.A. 2002. Selective feeding in the hawksbill turtle, an important predator in coral reef ecosystems. Marine Ecology Progress Series 245: 249-258.

Losey, G.S., Balazs, G.H., Privitera, L.A. 1994. Cleaning symbiosis between the wrasse, Thalassoma duperry, and the green turtle, *Chelonia mydas*. Copeia 5: 684-690.

Lutz, P.L. and J.A. Musick (Eds.). 1997. Biology of sea turtles. Vol. I. USA: CRC Press.

Lutz, P.L., J.A. Musick and J. Wyneken. 2002. Biology of sea turtles. Vol. II. USA: CRC Press.

Lynam, C.P., et al. 2006. Purcell, J.E. and Arai, M.N. 2001. Interactions of pelagic cnidarians and ctenophores with fish: A review. *Hydrobiologia* 451: 27-44.

Marques M., Rene. FAO Species Catalogue, Vol. 11. An Annotated and Illustrated Catalogue of Sea Turtles of the World. Rome: FAO, 1990.

Meylan, A. 1988. Spongivory in hawksbill turtles: A diet of glass. *Science* 239(4838): 393-395.

Mukherjee, N., A. Sridhar, M. Menon, S. Rodriguez and K. Shanker. 2008a. Policy brief: Bioshields.

Mukherjee, N., F. Dahdouh-Guebas and K. Shanker. 2008b. Beyond the tsunami: Current status of mangroves in Kerala and Tamil Nadu, India, with regard to vegetation, community perceptions and policy. UNDP/UNTRS, Chennai and ATREE, Bangalore, India.pp. 28.

Musick, J.A. and Limpus, C.J. 1997. Habitat utilization and migration in juvenile sea turtles. In: P. L. Lutz & J. A. Musick (Eds).

Namboothri, N., A. Swaminathan, M. Chandi and K. Shanker. 2011. Marine turtles of the Andaman and Nicobar Islands. Report submitted to the Andaman and Nicobar Forest Department. pp. 11.

Namboothri, N., D. Subramanian, A. Sridhar, S. Rodriguez, M. Menon and K. Shanker. 2008a.

Namboothri, N., D. Subramanian, B. Muthuraman, A. Sridhar, S. Rodriguez and K. Shanker. 2008b. Beyond the tsunami: Coastal sand dunes of Tamil Nadu, India - An overview. UNDP/UNTRS, Chennai and ATREE, Bangalore, India.pp. 56.

National Marine Fisheries Service and US Fish and Wildlife Service. Recovery Plan for Leatherback Turtles in the US Caribbean, Atlantic, and gulf of Mexico. Washington, D.C.: National Marine Fisheries Service, 1992.

National Marine Fisheries Service and US Fish and Wildlife Service.Recovery Plan for U.S. Population of Atlantic Green Turtle. Washington, D.C.: National Marine Fisheries Service, 1991.

Nicholls, Richard E. The Running Press Book of Turtles. Philadelphia, Pennsylvania: Running Press, 1977.

Obst, Fritz Jurgen. Turtles, Tortoises and Terrapins. New York: St. Martin's Press, 1986.

Palli, L., Musick, J.A. and Wyneken, J. (Eds.) The Biology of Sea Turtles Volume II. CRC Press, Boca Raton, Florida (USA). pp. 259-273.

Pandav, B. 2000. Post cyclone situation in coastal Orissa with special reference to marine turtle conservation. A GOI-UNDP Sea Turtle Project Report.Wildlife Institute of India, Dehradun. pp. 14.

Pandav, B. 2000.Conservation and management of olive ridley sea turtles on the Orissa coast. PhD thesis. UtkalUniversity,Bhubaneswar, India.Pandav, B. and B.C. Choudhury. 2000.Conservation and management of olive ridley sea turtle (*Lepidochelys olivacea*) in Orissa.FinalReport. Wildlife Institute of India, Dehradun.

Pandav, B., B.C. Choudhury and C.S. Kar. 1994. A status survey of olive ridley sea turtle (*Lepidochelys olivacea*) and its nesting habitats along the Orissa coast, India. Wildlife Institute of India.

Policy Brief: Sand Dunes. UNDP/UNTRS, Chennai and ATREE, Bangalore, India. pp. 12.

Pritchard, D. Peter C.H. Encyclopedia of Turtles. Neptune, New Jersey: T.F.H. Publications Inc., Ltd., 1979.

Prusty, B.G., R.K. Sahoo and S.D. Mehta. 2000. Natural causeslead to mass exodus of olive ridley turtles from Ekakulanasi,Orissa, India: a need for identification of alternate sites. In: Sea Turtles of the Indo-Pacific: Research, Conservation and Management (eds. N. Pilcher and G. Ismail) pp. 189–197. ASEAN Academic Press, London.

Purcell, J.E., Shin-ichi, U., Wen-Tseng, L. 2007. Anthropogenic causes of jellyfish blooms and their direct consequences for humans: *A review. Mar. Ecol. Prog.*

Rees, C. 1990. A guide to development in urban and coastal areas. A project of Asian Wetland Bureau funded by the Australian International Development Assistance Bureau (AIDAB).

Rodriguez, S., D. Subramanian, A. Sridhar, M. Menon and K. Shanker. Policy brief: Sea walls. UNDP/UNTRS, Chennai and ATREE, Bangalore, India.pp. 8.

Rudlow, Anne and Jack Rudloe."Sea Turtles: In a Race for Survival."National Geographic. 185(2), February, 1994.

Sazima, I., Grossman, A., Sazima, C. 2004. Hawksbill turtles visit mustached barbers: cleaning symbiosis between Eretmochelys imbricata and the shrimp, Stenopus Press, Washington, DC. pp. 235-254.

Schofield, G., Katselidis, K.A., Dimopoulos, P., Pantis, J.D., Hays, G.C. 2006. Behaviour analysis of the loggerhead sea turtle Caretta caretta from direct in-water observation. *Endangered Species Research* 2: 71-79.

Shanker, K. and B.C. Choudhury. (Eds.). 2006. Marine turtles of the Indian subcontinent. 416 p. Universities Press (India) Private Ltd.

Shanker, K. and H.V. Andrews. 2006. Monitoring and networking for sea turtle conservation in India: A UNEP CMS project report. Centre for Herpetology/ Madras Crocodile Bank Trust, Tamil Nadu.pp. 78.

Shanker, K., B.C. Choudhury and C.S. Kar. 2010. Census techniques for arribadas: Monitoring oliveridley sea turtles in Orissa. Ashoka Trust for Research in Ecology and the Environment, Bangalore, India. (Original edition 2007).Pp. 10.

Shanker, K., N. Namboothri, S. Rodriguez and A. Sridhar (Eds.). 2008. Beyond the tsunami: Social, ecological and policy analyses of coastal and marine systems on the mainland coast of India.

Shenoy, S., T. Berlie, N. Namboothri and K. Shanker. 2010. Building a network for conservation of marine turtles in India. Project report submitted to the USFWS. Ashoka Trust for Research in Ecology and the Environment, Bangalore.pp 57.

Sunderraj, W.S.F., V. Vijay Kumar, J. Joshua, S. Serebiah, I.L. Patel and A. Saravana Kumar. 2002. Status of the breeding population of sea turtles along the Gujarat coast. A GOI-UNDP Sea Turtle Project Report.Gujarat Institute of Desert Ecology, Bhuj, India.

Tripathy, B. 2004. A study of the offshore distribution of olive ridley turtles (Lepidochelysolivacea) in the coastal waters of Rushikulya rookery along the Orissa coast, India. Wildlife Conservation Society – India Programme Small Grant, Centre For Wildlife Studies, Bangalore, India.

Tripathy, B. and B.C. Choudhury. 2001. Sea turtles and their nesting beaches along the Andhra Pradesh coast, India: A status survey. A GOI-UNDP Sea Turtle Project Report. Wildlife Institute of India, Dehradun.

Tripathy, B., B.C. Choudhury and K. Shanker. 2002. A status survey of sea turtles and their nesting habitats in Lakshadweep islands, India. A GOI-UNDP Sea Turtle Project Report. Wildlife Institute of India, Dehradun.

Tripathy, B., R.S. Kumar, B.C. Choudhury, K. Sivakumar and A.K. Nayak. 2009. Compilation of research information on biological and behavioural aspects of olive ridley turtles along the Orissa coast of India – A bibliographical review for identifying gap areas of research. Wildlife Institute of India, Dehra Dun. Available at: http://www.wii.gov.in/index.php?option=com_content and view=article and id=98 and Itemid=299.

UNDP/UNTRS, Chennai and ATREE, Bangalore, India.pp. 397.

UNDP/UNTRS, Chennai and ATREE, Bangalore, India.pp. 8.

Van Meter, Victoria Brook.Florida's Sea Turtles. Atlanta, Georgia: Applied Biology, Inc. (for the Florida Power and Light Company, Miami Florida), 1983.

Witherington, B.E. and R.E. Martin. 2003. Understanding, assessing, and resolving light-pollution problems on sea turtle nesting beaches. 3rd ed. rev. Florida Marine Research Institute Technical Report TR-2. pp. 73.

Witzell, W.N. Synopis of Biological Data on the Hawksbill Turtle. Rome: Food and Agriculture Organization of the United Nations, 1983.

Wolke, R.E. and A. George. 1981. Sea turtle necropsy manual. NOAA Technical Memorandum NMFS-SEFC-24.

Wyneken, J. 2001. The anatomy of sea turtles. U.S. Department of Commerce NOAA Technical Memorandum NMFS-SEFSC-470. pp. 17.

Chapter 12

Present Status and Conservation of Monitor Lizard

Bindu Sharma

Department of Zoology
D.N. College, Meerut, U.P.

ABSTRACT

The Varanidae as a group of mostly big-growing lizards and comprising the largest living lizards of the world, should be expected to be more or less well known. However, despite the remarkable body size of many species, and despite their popularity and their meaning in the international reptile leather trade, not even the number of species existing on the earth is yet known. New species are continuously described, mainly from two sources:

1. *More commonly by re-evaluating and re-estimating the taxonomic rank of formerly described subspecies, due to new methodological approaches;*
2. *Less commonly by the discovery of new species from remote areas not seen before by herpetologists.*

Keywords: *Conservation, Monitor Lizard, Habitat.*

In classical monograph of the family Varanidae, Robert Mertens (1942) listed 24 different species. Of these, he believed 14 to be monotypic, and 10 to be polytypic (with a total of 31 subspecies) which made up 45 "species-group names" (ICZN) considered to be valid. In his subseqent checklist in the famous series "Das Tierreich" (1963) he listed 30 species, 18 of them being monotypic and 12 polytypic, with a total of 57 species group names regarded as valid. The next review of all varanid taxa known at that time was given by Bohme & Ziegler (1997) in the course of their taxonomic revision of the family Varanidae which was mainly based on genitalmorphological characters. They distinguished between 47 species: 37 monotypic and 10 polytypic ones, and a total of 67 species-group names regarded valid. Since this last summary and update of varanid taxonomy, even more new and spectacular discoveries of big-growing monitor lizard species have been made. Also, taxonomic rearrangements of former nominal subspecies to full species rank led again to an increase in the number of varanid species currently known, but also to a further reduction of the percentage of polytypic species as compared with monotypic ones.

The family of monitor lizards includes the largest lizards now in existence; several species reach 5ft or more. The largest is the Komodo dragon at 20 ft long, and the smallest is the 8-in. short-tailed monitor of Australia. The word monitor originally meant 'one who admonishes others about their conduct'. It was applied to this conduct'. It was applied to these lizards because of an error in translation. The Arabic name for the lizards is waran which is very similar to the German warnen meaning 'warning'. As a result these lizards became known as 'warning lizards'. In Malaya and Australia monitors are called iguanas or 'goannas' although the iguanas form a separate family.

Monitors live in the warmer parts of the Old World, including the whole of Africa excluding Madagascar, Asia from Arabia to southern China down through Southeast Asia to the East Indies and Australia. Unlike other lizard families where some species carry ornate crests, spines and frills, the monitors show very little variation in form. The snout, neck, body and tail are all long and slender and their eyes are prominent, making the lizard seem long, sinuous and alert. Monitors have several features in common with snakes although they are not as closely related to them as the earless monitors. Both snakes and monitors have long, forked tongues and have lost the ability to shed their tails.

The largest monitor, excluding the Komodo dragon, is the two-banded or water monitor, at 8–9 ft. *Varanus giganteus*, the perenty, reaches 7 ft, and the Nile monitor 6 ft. The Nile monitor ranges throughout Africa from the Upper Nile south to the Cape and the water monitor lives in Malaya.

The Nile monitor is brownish or greenish grey above with darker markings and yellowish spots which are lost with age. Underneath the skin is yellowish with dark bands. The Indian monitor is light to dark brown, sometimes with scattered light and dark scales on its back. Its underparts are dirty white with speckling. Young monitors, and adults, just after they have shed their skins, are more brightly colored. Young Indian monitors are orange to light brown with alternating yellow and black bands across the back.

A myth, that is extraordinarily common considering its stupidity, is that an animal "will grow to the size of its surroundings, and then stop"! This, of course, is utter nonsense. A healthy reptile never stops growing, from the day it is laid to the day it dies. Many monitor lizards spend most of the day fast asleep, and may not initially appear to very active animals. However when they do move they tend to cover a lot of distance. Monitors are large lizards which can get quite nasty if brought to bay. At first they inflate their bodies and hiss. Then they attempt deter any attack by violently lashing the tail like a whip. Finally they may attack by grabbing their adversary with their powerful jaws and clawing with their feet. Dogs often come off worst in such an encounter and a large monitor is a formidable opponent for a man. They may spend much of their time, however, basking in the sun and when alarmed they can run very rapidly with the tail held in a characteristic curve with the tip just off the ground. The Nile monitor can outpace a man, especially in thick cover. Except during cold seasons monitors are extremely active animals.

Each species of monitor has a fairly restricted habitat. The perenty, the largest of the Australian monitors, lives among rocky outcrops in the deserts. The desert monitor lives in similar habitats in North Africa and western Asia. Most monitors can climb trees but some forest-dwelling species are completely at home in trees. The Nile monitor and water monitor have flattened tails that are used for swimming and the water and the water monitor has nostrils near the tip of its snout which allows it to breathe when almost completely submerged. When disturbed, monitors make for their natural home. Nile monitors usually bask on a branch or rock from which they can drop straight into the water. When surprised away from water they usually head for the bank but will sometimes hide in a tree or hole. The Indian monitor and the desert monitor also hide in holes in the ground while the forest-living Bornean rough-necked monitor takes refuge in trees.

Like snakes, monitors swallow their prey whole rather than chew it as do iguanas and other lizards. A water monitor, for instance, has been known to swallow a 6 in. turtle whole. Snakes and monitors have a strong bony roof of the mouth which protects the brain form being damaged by the passage of large mouthfuls.

Monitors are carnivorous, eating a wide variety of animals, as well as carrion. The Indian monitor eats palm squirrels, musk shrews, lizards, snakes and invertebrates and the water monitor feeds largely on fish. The Nile monitor eats mainly insects, snails, crabs and other invertebrates and also catches small birds and mammals. In South Africa, the Nile monitor digs rain frogs out of their burrows, ignoring the poisonous secretions that deter other predators. All monitors have a fondness for eggs and as a result are often persecuted by farmers for taking the eggs of poultry. The Nile monitor is said to steal the eggs from crocodiles' nests even though these are guarded by the mother.

Monitors usually lay their eggs in nests in the ground or in hollow trees. The Indian monitor digs a hole about 1 ft deep with its forelegs then turns and backs in to lay its eggs. After laying it fills the nest hole by breaking down the sides with its snout and claws and tamping down the soil. In South Africa, Nile monitors lay their eggs in termite nests. They tear holes in the sides of the nests, lay their eggs in an Enlarged chamber in the centre then leave termites to repair the damage and so cover the eggs. Between 16 and 34 white leathery eggs are laid. The termites' nest provides heat for incubation and safety from enemies until the moniters hatch 5 months later.

Conservation Status

Monitor lizards of the family Varanidae represent an ancient group of anguimorph reptiles. Next to crocodiles and pythons, they comprise the largest living poikilothermic predators on our planet. Monitor lizards inhabit Africa, the Arabian Peninsula, the Middle East as well as South and Southeast Asia. India has a great diversity of herpeto fauna belonging to some 137 genera and 26 familes which is close to 8 per cent of the world's fauna. Many of India's herbs are unique, endemic or of relict form. Man has extensively exploited lizards including monitor lizards for food and his social needs. In India Monitor Lizards are eaten and they are also a source of leather. Live monitors are used in India in fertility rites, or serpent festivals. Over-hunting combined with the problem of habitat loss, has resulted these species being pushed to endangered status.

Some species of monitor lizard popular in the wildlife trade belong to species restricted to tiny islands in South East Asia. Very few of these animals ever reproduce in captivity and the actual trade vastly exceeds the declared trade. Virtually nothing is known of the population status or ecology of any of these species. Captive breeding of these species is restricted to very rare, isolated events. Some of the species occur only on extremely small islands, and unlikely to occur in large numbers. Many people consider the pet trade to be a significant threat to the conservation of these species but, because they are virtually unstudied and levels of exploitation cannot be documented, very little hard evidence exists.

In India four kinds of monitors occur. They are Bengal monitor lizazd, Varanus bengalensis; water monitor lizard, Varanus salvator; desert monitor lizard, Varanus griseus and yellow monitor lizard, Varanus flavescens. Of the above four species, the Bengal monitor lizards Varanus bengalensis is very common. Detailed information on the present status and distribution of these monitors is lacking but many of above listed species has been classified as vulnerable in the IUCN Red List (International Union for Conservation of Nature) since 1996. Globally, monitor lizards and their products are important goods in the world-spanning trade with live pets and reptile leather. Consequently, all members of the genus Varanus are protected by the Convention on International Trade in Endangered Species of Wild Fauna and Flora (CITES).

All the species reported were facing threat due to human activity. These animals were still hunted for illegal export of skin through border countries like Bengladesh, Nepal and Pakistan. They were also extensively used within the country to make musical instruments, extract oil for medicines for body ailments and meat for food. In some places, they were killed thinking that these animals as poisonous.

Conservation Goals

- ✰ Surveys of poorly known species (*V. griseus, V. flavescens*) to assess distribution and status
- ✰ Protection of critical habitats for wild populations.
- ✰ Prevention of illegal hunting and trade.
- ✰ Awareness programmes to revive support for monitor lizard conservation and
- ✰ Establishment of sustainable use programmes for common species.

References

Böhme, W., 1991. The identity of *Varanus gouldii* (Gray, 1838), and the nomenclature of the *V. gouldii* species complex. - In Böhme, W. & H.-G. Horn (eds.): Advances in monitor research. Mertensiella 2: 38-41.

Böhme, W. & T. Ziegler, 1997. On the synonymy and taxonomy of the Bengal monitor lizard, *Varanus bengalensis* (Daudin, 1802) complex (Sauria: Varanidae). *Amphibia-Reptilia* 18 (2): 207- 211.

Böhme (2003). Checklist of the living monitor lizards of the world. *Zool.* Verh. Leiden.

Branch, W. R., 1982. Hemipenial morphology of platynotan lizards. *J. Herpetol.* 16: 16-38.

International Commission of Zoological Nomenclature, 2000. International Code of Zoological Nomenclature. London (ICZN), 4th ed., 306 pp.

King, D., King, M. & P. Baverstock, 1991. A new phylogeny of the Varanidae. - In Böhme, W. & H.-G. Horn (eds.): Advances in monitor research. Mertensiella 2: 211-219.

King, D., Fuller, S. & P. Baverstock, 1999. The biogeographic origins of varanid lizards. - In Horn, H.-G. & W. Böhme (eds.): Advances in monitor research II. Mertensiella 11: 43-49.

Koch, A., Ziegler, T., Boehme, W., Arida, E., & Auliya, M. (2013). Pressing Problems: Distribution, threats, and conservation status of the monitor lizards (Varanidae: Varanus spp.) of Southeast Asia and the Indo-Australian Archipelago. *Herpetological Conservation and Biology, 8*, 1-62.

Lauprasert, K. & K. Thirakupt, 2001. Species diversity, distribution and proposed status of monitor lizards (family Varanidae) in southern Thailand. Nat. Hist. *J. Chulalongkorn Univ.* 1 (1): 39-46.

Sprackland, R. G., 1991. Taxonomic review of the *Varanus prasinus* group with descrition of two new species. Mem. Qld. Museum 30 (3): 561-576.

Pianka, E. R., King, D., & King, R. A. (2004). *Varanoid lizards of the world*. Indiana University Press.

Chapter 13

World's Weirdest Blood Squirting Lizard: The Horned Toad

Bindu Sharma

Department of Zoology
D.N. College, Meerut, U.P.

ABSTRACT

Horned lizards is a genus, Phrynosoma of lizards which are type genus of the family "Phrynosomatidae. The horned lizard is popularly called a "horned toad", "horny toad" or "horned frog", yet it is neither a frog. The popular names come from the lizard's rounded body and blunt snout, which make it resemble a toad or frog (Phrynosoma literally means "toad-bodied"). The spines on its back and sides are made from scales, whereas the horns on the heads are true horns (i.e. they have a bony core). Of 15 species of horned lizards in North America, eight are native to United States.

Keywords: *Conservation, Habitat, Horned toad.*

The horned toad is not a toad as the name suggests but a lizard with the face of a toad. If it were larger it could be mistaken for a prehistoric reptile. As it is, its main claim to fame is its alleged ability to squirt blood out of its eyes.

It can measure up to 5 in. long, and has a squat, flattened, almost circular body, short tail. The head is ornamented with backwardly directed spines, the so-called horns, and the back is covered with smaller spines. Some have been called short-horned, because the head spines are not prominent others are called long-horned. In both types the body is covered in small scales, as is usual in lizards, but there is larger horn like scales as well. Usually the edges of the body are ornamented with large flat scales.

There are a dozen species of horned toads or horned lizards ranging from just over the Canadian border southwards through the western United States to Mexico.

The most widely distributed is the Texas horned toad, from Nebraska in the north to Chihuahua and Sonora in Mexico.

The most common horned lizard in the western deserts is aptly named the **desert horned lizard** (*Phrynosoma platyrhinos*) consisting of two subspecies: the **northern** (*P. platyrhinos*) which inhabits the Great Basin Desert, and the **southern** (*P.p. calidiarum*) which inhabits the Sonoran and Mojave Deserts including a finger of the east coast of northern Baja California.

Five other species of horned lizards inhabit the North American deserts:

1. **Roundtail Horned Lizard** (*Phrynosoma modestum*) Chihuahuan
2. **Flattail Horned Lizard** (*Phrynosoma m'calli)* Sonoran
3. **Regal Horned Lizard** *(Phrynosoma solare)* Sonoran
4. **Texas Horned Lizard** *(Phrynosoma cornutum)* Chihuahuan
5. **Shorthorned Lizard** *(Phrynosoma douglassi)* Great Basin

The **coast horned lizard** (*Phrynosoma coronatum*), which is found in coastal and cismontane California, crosses to the east side of the Baja Peninsula, actually making contact with the desert horned lizard in the vicinity of Bahía de Los Angeles. The two live sympatrically for a distance along the east coast. As the coastal form has taken over the balance of the peninsula, it could also be considered a desert form since it does live in the deserts of Baja right down to Cabo San Lucas. There is, however, some conjecture about the taxonomy of the **Baja horned lizards** starting with the ones found directly south of the San Diego subspecies, and they may actually be one or more different species/subspecies related to the coast horned lizard.

These lizards live in deserts and semi-desert sandy country, from low-lying ground to 10,000 ft altitude. They drink dew and hunt insects, particularly ants. A horned toad moves slowly forward towards its prey and, when close enough, stops and bends its head slightly towards it. Then it shoots out a thick tongue and in a flash carries the insect back into its mouth. As the day wanes the horned toad buries itself in the warm sand. It pushes its blunt snout into the sand and by wriggling strenuously makes a furrow in which it lies half buried, or with only the top of the head showing. As autumn approaches the horned toad spends more and more time buried and in winter it buries itself deeper and goes into a torpid state.

Most horned lizards are egg-layers. Between April and July the female digs a hole 6 in. deep in the sand. She does this with her forefeet, pushing the sand back with her hind feet. She lays up to 30 yellowish-white, oval, tough-shelled eggs each ½ in long, in the hole. These she covers with sand and leaves. The eggs hatch up to 90 days later, the time varying with the species. A few horned lizards bear live young, up to 30 at a time. These are form eggs that hatch just before being laid.

Horned toads have few enemies because of their armor and the camouflage effect of their colors. Snakes sometimes eat them and often pay for this with their lives since the horns of the lizard may penetrate the wall of the snake's gullet. They are also well camouflaged, for not only do the lizards bury themselves in the sand, but the mottled colors on their bodies tend to take on the color or pattern of the sand or gravel o which they are living.

Horned toads are said to squirt blood from their eyes when alarmed. Yet there are many who have kept the lizards as pets or are used to seeing them around who say this is untrue. It is clear form all the evidence that they do this do this but in a curiously erratic way. Raymond L Ditmars, American specialist in the study of reptiles; and author of The Book of Living Reptiles (1936), handled several hundred horned toads before he saw one squirt blood from its eyes. Another American specialist, Winton, writing in 1914, tells how one of his students stooped to pick up horned toads and received a splash of blood on his hand which spread in a fan-shaped smear form the second joint of the index finger to the wrist. Winton saw three toads do this. All were males and all were sloughing their skins at the time. In each, the eye from which the blood was ejected showed a small quantity of clotted blood in the back of the cornea, in which the blood vessels were swollen although the cornea itself remained intact. So it seems one person may handle three of the lizards and see the blood-letting three times, while another may handle hundreds over periods of years and never see it.

The usual explanation is that this action is defensive, and that the blood is an irritant to the eyes of small mammals. Another suggestion is that it may be connected with the breeding season, although nobody is very clear how. A third suggestion is that it may be due to a parasite, and a fourth is that it may be a secondary use acquired by relatively few individuals.

This habit is not unique, however, only particularly spectacular. It is like the 'blood-spitting' of the dwarf boas of the West Indies.

Regardless of where they occur, there is a similarity in their habits. In the fall, they hibernate by burying themselves in the sand. They emerge in the spring when the sun's rays have reached a certain temperature. The first few hours of the day are spent basking, usually flattened against a rock or on slanting soil, so their back is exposed to the sun. At times, while warming up, they may flatten and tilt their bodies toward the sun to obtain maximum radiation.

As soon as their body temperature rises to a specific degree, they begin foraging for food. As the heat of the day increases, they become more active. They feed on slow-moving, ground-dwelling insects, spiders, sow bugs, an occasional tick and even items as large as the butterfly and sphynx moth larvae.

Ants seem to be their major food source. They do not pursue their victim hastily, like some lizards, but poise over it and methodically take it, in toad-like fashion, with a flick of their long, sticky tongue. The toad-like action ceases if disturbed, for they will flee as rapidly as a startled mouse.

They stick their nose in the sand like the blade of a plow and wriggle forward to create a short furrow. After flattening the body, they use the spiny border of the sides in a shovel-like fashion to scoop and dig their way into the sand. Sometimes they bury themselves 3 or 4 inches deep, and other times they just leave the top of the head and eyes exposed.

Their coloration is such that they blend readily into their surroundings, making them difficult to find. However, when found partially covered with sand, they are rather easily captured. Their defense mechanisms are quite limited. When caught by hand, they may distend their bodies by filling their lungs with air and twist their head in a futile attempt to scratch you with their horns. On occasion they spurt blood from the corners of their eyes, which is startling, to say the least.

Conservation Status

There are a dozen species of horned toads or horned lizards ranging from just over the Canadian border southwards through the western United States to Mexico. The most widely distributed is the Texas horned toad, from Nebraska in the north to Chihuahua and Sonora in Mexico. Over-hunting combined with the problem of habitat loss, has resulted these species being pushed to endangered status. Detailed information on the present status and distribution of these horned toads is lacking but many of the species has been classified as vulnerable in the IUCN Red List (International Union for Conservation of Nature). Globally, consequently, all members of these are protected by the Convention on International Trade in Endangered Species of Wild Fauna and Flora (CITES). All the species reported were facing threat due to human activity. These animals were still hunted for illegal export of skin through border countries.

Some do diverge from the ant diet at certain times of the year such as the regal horned lizard, which gorges itself on tiny beetles and eschews ants altogether when the beetles are abundant. Also, the coast horned lizard can survive on inverts other than ants. But the flat tail and short horned, as well as the desert horned lizards, are closely tied to ants and will die if those are not supplied in quantity.

Habitat destruction and ant destruction have placed several species of horned lizards in danger. After all, the first thing people do when they move into the desert is kill the pesky ants, thereby depriving horned lizards of their only dependable diet. Taken from their native surroundings and offered an improper diet and an inadequate place to live, horned lizards soon die.

Conservation Goals

- ☆ Surveys of poorly known species to assess distribution and status
- ☆ Protection of critical habitats for wild populations.
- ☆ Prevention of illegal hunting and trade.
- ☆ Awareness programmes to revive support for Horned toad conservation and
- ☆ Establishment of sustainable use programmes for common species.

References

Hodges, W. L. (2004). Evolution of viviparity in horned lizards (*Phrynosoma*): testing the cold climate hypothesis. *Journal of Evolutionary Biology, 17* (6), 1230-1237.

Leaché, A. D., Koo, M. S., Spencer, C. L., Papenfuss, T. J., Fisher, R. N., & McGuire, J. A. (2009). Quantifying ecological, morphological, and genetic variation to delimit species in the coast horned lizard species complex (*Phrynosoma*). *Proceedings of the National Academy of Sciences, 106* (30), 12418-12423.

Lizard, G. S. H. Recovery Strategy for the Greater Short-horned Lizard (*Phrynosoma hernandesi*) in Canada.

McIntyre, N. E. (2003). Effects of Conservation Reserve Programme seeding regime on harvester ants (*Pogonomyrmex*), with implications for the threatened Texas horned lizard (*Phrynosoma cornutum*). *The Southwestern Naturalist*, *48*(2), 274-277.

Moeller, B. A., Hellgren, E. C., Ruthven III, D. C., Kazmaier, R. T., & Synatzske, D. R. (2005). Temporal differences in activity patterns of male and female Texas horned lizards (*Phrynosoma cornutum*) in southern Texas. *Journal of Herpetology*, *39*(2), 336-339.

Whiting, M. J., Dixon, J. R., & Murray, R. C. (1993). Spatial distribution of a population of Texas horned lizards (*Phrynosoma cornutum: Phrynosomatidae*) relative to habitat and prey. *The Southwestern Naturalist*, 150-154.

Winton, W. M. (1914). A Note on Distinction of the Sexes in Phrynosoma. *Science*, 311-312.Ditmars, R. L. (1936). *The Book of Living Reptiles*. JB Lippincott Company.

Chapter 14

The Wildlife (Protection) Act, 1972

Hemlata Wankhede[1] and Yogesh Reddy[2]

[1]Govt Pre IAS Training Center,
Aurangabad, M.S.
P.G. Dept.of Zoology,
[2]Deogiri College, Aurangabad, M.S.

ABSTRACT

Wild animals means any non-domesticated animals found wild in nature. As pre act it includes Wild animals,birds and plants and for matters connected there thereto.The Wildlife of India is unique in richness of its variety.Himalayas in north to the virgin forest of south and from the mangrove swamps in east to the sprawling deserts of the west Indian tiger the most famous animal of the Indian. Since 1972 it has also became India's National Animal. Due to unrestricted killing of wild animals their no became very less in the forest hence to save them from extinction the Govt of India framed the wildlife act for the conservation of the wild species.

Keywords: *Wildlife act, Endangered species, Authorities.*

Introduction

The wildlife animals of India including Indian Lion, Indian trigger, Indian rhino, Indian elephant, Indian crocodiles, Great Indian bustard, Common pea fowl, Hornbills are going to became extint. The Rule and regulation to wildlife protection become the talk of the town and hug the lime light when a filmstar is arrested for hunting a wild blackbuck or when a dinner party with exotic dishes. There are many cases which come across regarding seizure of tiger skin, elephant tusks, rare butterflies, star turtles and rare birds at the time of the smuggling (Singhar, 2002). The Wildlife Law in India has a Long History.The Earliest codified law can be traced back to the Thirtheen century BC, when emperor Ashoka Inacted a Law in the matter of prevention of wildlife and environment (Joshi *et al.*, 1998). In the pre-constitutional period, there were a few legislations which are enacted to protect the wildlife from exploitation. The Cattle Trespass Act, 1871; The Elephants Preservation Act, 1879; some sections of Indian Penal Code, 1860. The first codified law, The Wild Bird Protection Act 1887 was enacted by the British. The British Government passed The Wild Birds and Animals Protection Act, 1912. The Indian Forest Act, 1927 are some pre-constitutional enactments on wildlife protection and realised that these laws were not adequate to protect the wildlife. Then the Government of India brought out a comprehensive legislation, the wildlife (Protection) Act, 1972 (WPA). WPA provides the basic frame work to ensure the safety and well being of wild animal and plants (Singhar,2002).

The Wildlife Protection Act 1972 defines wildlife includes any animal, bees butterflies, crustacean, fish and moths; and aquatic their eggs or land vegetation which forms part of any habitat. So the meaning of the wildlife in this Act is very wide and inclusive of all kinds of flora and fauna. The Wildlife Protection Act, 1972, provides for protection to listed species of flora and fauna and establishes a network of ecologically-important protected areas. The Act consists of 60 Sections and VI Schedules- divided into Eight Chapters. The Wildlife Protection Act, 1972 empowers the central and state governments to declare any area a wildlife sanctuary, national park or closed area. There is a blanket ban on carrying out any industrial activity inside these protected areas. It provides for authorities to administer and implement the Act; regulate the hunting of wild animals; protect specified plants, sanctuaries, national parks and closed areas; restrict trade or commerce in wild animals or animal articles; and miscellaneous matters. The Act prohibits hunting of animals except with permission of authorized officer when an animal has become dangerous to human life or property or as disabled or diseased as to be beyond recovery.

Territorial Jurisdiction of Wildlife (Protection) Act, 1972

Wildlife (Protection) Act, 1972 has been accepted and adopted by all the states except Jammu and Kashmir. This is the first comprehensive legislation relating to protection of wildlife was passed by the Parliament and it was assented by the President on 9th September, 1972 and came to be known as The Wildlife (Protection) Act, 1972.

Authorities Constituted Under Wildlife (Protection) Act

As per the Sec. 3 of the Act, the Central Government may appoint a Director of Wildlife Preservation, Assistant Directors of Wildlife Preservation and such other officers and employees as may be necessary. As per the Sec. 4, the State Government may, for the purpose of this Act, appoint Chief Wildlife Warden, Wildlife Warden, Honorary Wildlife Wardens and other officers and employees as may be necessary. As per Sec. 6, the State Governments and the Administrators in Union Territories shall constitute a Wildlife Advisory Board.

The Members of the Wildlife Advisory Board

It shall consist of the Minister in charge of Forests in the State or Union territory as the Chairman. If there is no such minister, then the Chief Secretary will be the Chairman of the Board. The other members are, two members of the State

Legislature or Legislature of Union Territory; Secretary to the state government or the government of the union territory, in charge of forests; the Forest Officer in charge of the State Forest Department; an officer nominated by the Director of Wildlife Preservation; Chief wildlife warden; officers of the state government not exceeding 5; and such other persons, not exceeding 10, who in the opinion of the state government, are interested in the protection of wildlife, including the representatives of tribals not exceeding 3.

Duties of Wildlife Advisory Board

The Wildlife Advisory Board mainly constituted to advise the state government in the following matters.

1. In the selection of areas to be declared as Sanctuaries, National Parks and Closed areas and the administration thereof;
2. In formulation of the policy for protection and conservation of wildlife and specified plants;
3. In any matter relating to the amendment of any schedule;
4. In relation to the measure to be taken for harmonizing the needs of the tribals and other dwellers of the forests with the protection and conservation of wildlife;
5. In any other matter connected with the protection of wildlife which may be referred to it by the state government.

List of Mammals Under Wildlife Protection Act 1972

Andaman wild pig (*Sus andamanensis*), Bharal (*Ovis nahura*) Binturong (*Arctictis binturong*), Blackbuck (*Antilope cervicapra*) Brown-antlered deer or thami, (*Cervus eldi*), Himalayan brown bear (*Ursus arctos*), Capped langur (*Presbytis pileatus*), Caracal (Felis caracal), Catacean specils Cheetah (*Acinonyx jubatus*) Chinese pangolin (*Manis pentadactyla*), Chinkara or Indian gazelle (*Gazella gazella bennetti*), Clouded leopard (*Neofelis nebulosa*), Crab-eating macaque (Macaca irus umbrosa), Desert cat (*Felis Libyca*) Desert fox (*Vulpes bucopus*), Dugong (*Dugong dugong*), Fishing cat (*Felis viverrina*), Four-horned antelope (*Tetraceros quadricorins*), Ermine (*Mustela erminea*), Gaur or Indian bison (*Bos gaurus*), Golden cat (*Felis termmincki*), Golden lungur (*Presbytis geei*), Giant squirrel (*Ratufa macroura*), Himalayan ibex (*Capra ibex*), Himalayan tahr (*Hermitragus jemlahicus*), Hispid hare (*caprolagus hispidus*), Hog badger (*Arctonyx collaris*), Hoolock gibbon (*Hylobates hoolock*), Indian elephant (*Elephas maximus*), India Lion (*Panthera leo persica*), Indian wolf ass (*Equus hemionous Khur*), Indian wolf (*Canis lupus pallipes*), Kashmir stag (*cervus elaphus hanglu*), Leaf monkey (*Presbytis phayrei*), Leopard cat (*Felis bengalensis*), Leopard cat (*Felis benegalensis*), Lessor or red panda (*Ailurus fulgens*), Lion-tailed macaque (*Macaca silenus*), Loris (*Loris tardigradus*), Little Indian porpoise (*Neomeris*), Lynx (*Felis lynx isabellinus*) Malabar civet (*Viverra megaspila*), Malay or sun bear (Helarctos malayanus) Marbled cat (*Felis marmorata*), Markhor (*Capra falconeri*), Mouse deer (*Tragulus meminna*) Musk deer (*Moschus moschiferus*) Nilgiri langur (*Presbytis johni*), Nilgiri tahr (*Hemitragus hylocrius*), Nayan or great Tibetan sheep (*Ovis ammon hodgsoni*), Pallas cat (*Felis manul*) Pangolin (Manis crassicaudata), Pygmy hog (*Sus salvanius*) Ratel (*Mellivora capensis*), Rhinoceros (*Rhinoceros unicornis*) Rusty-spotted cat (*Felis rubiginosa*), Serow (*Cparicorns sumatraensis*), Clawless otter (*Aonyx cinerea*) Sloth bear (*Melursus ursinus*) Slow loris (*Nyeticebus coveang*), Small Travancore flying squirrel (*Petinomys fuscopapillus*), Snow leopard (*Panthera uncia*) Snubfin dolphin (*Orcaella brevirostris*), Spotted linasng (*Panthera uncia*), Swamp deer (all sub-species of *Cerves duvauceli*), Takin or Mishmi takin (Budorcas taxicolor), Tibetan antelope or chiru (*Panthelopes hodgsoni*), Tibetan fox (*Vulpes ferilatus*), Tibetan gazella (*Procapra picticaudata*), Tibetan wild ass (*Equus heminous kiang*), Tiger (*Panthera tigris*), Urial or shapu (*Ovis vignel*), Wild buffalo (*Bubalus buballs*), Wild yak (*Bos grunniens*), Tibetan wolf (*Canis lupus chanco*).

List of Reptiles Under Wildlife Protection Act 1972

Agra Monitor Lizard [*Varanus griseus* (Daudin)], Audithia furtle (*Pelochelys bibroni*). Barred, oval, or yellow monitor lizard (*Varanus flavescens*) Crocodile (including estuarine or saltwater corcodile), (*Crocodile proosus* and *Crocodilus palustris*), Terrapin (*Batagur baska*), Eastern hill terrapin (*Melanochelys tricarinata*), Gharial (*Gavialis gangeticus*), Ganges soft-shelted turtle (*Trionyx gangeticus*), Golden gecko (*Caloductyloides aureus*), Green sea turtle (*Chelonia mydas*), Hawksbill turtle (*Eretmochelys imbricata imbricata*), Indian egg-eating snake (*Elachistodon westermanni*), Indian soft-shelled turtle (*Lissemys punctata*), Indian tent turtle (*Kachuga tecta tecta*), Kerla forest terrapin (*Hoesemys sylratiea*), Leathery turtle (*Dermochelys coriscea*), Loggerhead turtle (*Caretta caretta*), Oliveback loggerhead turtle (*Lepidochelys olivacea*), Peocock-marked soft-shelled turtle (*Trionx hurum*), Pythons (*Genus phython*), Sail terrapin (*Kachuga kachuga*), Spotted black terrapin (*Geoclemys hamiltoni*), Water Lizard (*Varanus salvator*).

List of Birds Under Wildlife Protection Act 1972

Andaman teal (*Anas gibberifrons albogularis*), Bengal florican (*Eupodotis bengalensis*), Black-necked crane (*Grus nigricollis*) Blood pheasants (*Lthaginis cruentus tibetanus, Lthaginis cruentus kuseri*), Cheer pheasant (*Catreus wallichi*), Eastern white stork (*Ciconia ciconia boyciana*), Forest spotted owlet (*Athene blewitti*), Frogmouths (*Genus batrachostomus*), Great Indian bustard (*Choriotis nigriceps*), Great Indian hornbill (*Buceros bicornis*), Hawks (family Accipitridae), Hooded crane (*Grus monacha*), Hornbills (*Ptiloaemus tickelli austeni, Aceros nipalensis, Rhyticeros undulatus ticehursti*), Houbara bustard (*Chlamydotis undulata*), Hume's bar-backed pheasant (*Syrmaticus humiae*), Indian pied hornbill (*Anthracoceros malabaricus*), Jerdon's courser (*Cursorius bitorquatus*), Lammergeier (*Gypaetus barbatus*), Largefalcons (*Falco peregrinus, Falco biarmicus, Falco chicquera*), Large whistling teal (*Anatidae*), Lesser florican (*Sypheotides indica*), Monal pheasants (*Lophophorus impejanus, Lophophorus sclateri*), Mountain quail (*Ophrysia supercilios*), Narcondam hornbill (*Rhyticeros (undulatus) narcondami*), Nicobar megapode (*Megapodius freycinet*), Nicobar pigeon (*Caloenas nicobarica pelewensis*), Osprey or Fish-eating eagle (*Pandion haliaetus*), Peacock pheasants (*Polyplectron bicalcaratum*), Peafowl (*Pavo cristatus*), Pink-headed duck (*Rhodonessa saryophyllacea*), Scalater's monal (*Lophophorus sclateri*), Siberian white crane (*Grus leucogeranus*), Tibetan snow cock (*Tetraogallus tibetanus*), Tragopan pheasants (*Tragopan melanocephalus, Tragopan blythii, Tragopan satyra, Tragopan temminckii*), White-bellied sea eagle (*Haliaetus leucogaster*), White-eared pheasant (*Crossoptilon crossoptilon, Platalea leucorodia*), White-winged wood duck (*Cairina scutalata*).

References

Joshi, A.N. Sardesai, R.S. Sharma, A. Mohammed and C.R. Babu (1998): Protection of Wild Genetic Resources of Plants and Microbes through Wildlife (Protection) Act, 1972. Handbook.

Singhar A. Samant (2002): Laws for Protection of Wildlife In India: Need for Awarness Toward Implementation and Effectiveness. Indian *Forester*, pp. 1113–1118.

Chapter 15

Wildlife Conservation and Management

Geeta Bansal[1], Vinay Gupta[2], Rashmi Gupta[1] and Seema Khan[1]

[1]Department of Zoology,
[2]Department of Chemistry,
D.N. (P.G), College, Meerut, U.P.

ABSTRACT

Wildlife conservation and management is the conservation, use and management of wild-animal populations and of the land necessary to support them to ensure that productivity and ecological balance are maintained in perpetuity, while social benefits are realized. Wildlife conservation and management in Canada is a reflection of legislation that can be divided into 2 groups: one concerning fisheries and marine mammals and one dealing with other forms of wildlife. Marine birds and mammals increasingly face the threat of offshore oil spills and general pollution of the oceans. The direct threat of uncontrolled harvest, so devastating in the 19th century, has been replaced by the indirect, insidious but permanent threats of environmental degradation that were characteristic of the 20th century. If society wishes to maintain wildlife in its variety and abundance in the 21st century, a place for wildlife must be maintained in land-use planning, environmental management and legal obligation.

The uses and value of wildlife to society are varied. Wildlife is one part of the equation which, together with vegetation and the nonliving environment, constitutes the "balance of nature," i.e., the set of complex natural processes on which human survival depends. Wildlife management draws on disciplines such as mathematics, chemistry, biology, ecology, climatology and geography to gain the best results.

Keywords: *Conservation, Management, Wildlife.*

Introduction

Wildlife comprises those forms of animal life that are not domesticated.

Individual members of wild species held tame in captivity are still considered "wildlife" as they are not genetically different from those remaining in a wild state. Human activity has become one of the most significant influences on the abundance and well-being of wildlife. Wildlife conservation aims to half the loss in the earth's biodiversity by taking into consideration ecological principles such as carrying capacity, disturbance and succession and environmental condition such as physical geography, pedology and hydrology with the aim of balancing the needs of wildlife with the needs of people.

History of Wildlife Conservation and Management

The first European explorers and settlers in North America found wildlife in abundance. This wealth was recognized as having immediate commercial value, with fisheries and the furtrade being the first widespread exploitive activities. As further exploration revealed the vast, sparsely populated expanse of land, it was believed that natural resources were unlimited; hence, there was no apparent need to practice conservation.

Wildlife, fish and timber were free for the taking for personal use, or could be converted into a monetary return. The practical result of this attitude became apparent in the latter half of the 19th century. Land development and uncontrolled harvest had extirpated them in that area by 1850. The great yuk disappeared in the early 1800s and the wild turkey by 1902 (Game birds), but only with the extinction of the once abundant passenger pigeon was there sufficient concern to cause the passage of wildlife conservation laws.

While concern and consequently protective legislation developed in eastern Canada, western and northern Canada were still held to be boundless frontiers. Wapiti rapidly diminished on the prairie; by 1890 only scattered, remnant populations remained throughout their former western range. In the 1820s bison teemed in millions, defying counting, across the North American plains. Their numbers remained significant into the late 1870s but, by 1885, they were almost gone.

Following Confederation and the constituation act, 1867, a move was made to develop wildlife conservation and protection areas. The first national park in Canada, Banaff national park (established in 1885), was not created to protect wildlife, although this became one of its significant functions. The first bird sanctuary in North America (and perhaps the Western Hemisphere) was created at Last Mountain Lake (Sask) in 1887; however, the declaration establishing it remained essentially unrecognized until after passage of the Canada Migratory Birds Convention Act in 1917.

The whooping crane population was protected from hunting, and its nesting and wintering areas have been safeguarded. Whitetailed deer now flourish in regrowth forests and in fringe agricultural areas. Wapiti, aided by transplanted populations, have regained strength in the mountain regions and in the localized areas of the prairie that are associated with parks.

Sea otters have responded to protection from commercial overharvest. Beaver have returned from dangerously low levels to become a problem species, causing damage where populations are not controlled. Through protection, northern tundra muskoxen have shown a dramatic recovery, as have prairie grassland pronghorn.

Wildlife Conservation and Management through Wildlife Laws

The Fisheries Act of Canada provides that all fisheries and marine mammals in Canadian waters are the responsibility of Canada's Department of Fisheries and Oceans (DFO). The DFO actively protects and manages marine mammals (seals, whales, walrus, etc) and offshore fisheries.

Marine fish and mammal populations are viewed primarily as a commercial resource. Legal harvesting seasons and quotas on lobster, salmon and other species reflect the continued monitoring of populations and harvest. Harvest management, together with pollution-control laws, is designed to maintain healthy, viable populations. The industry depends on the sustained yield of a quality product.

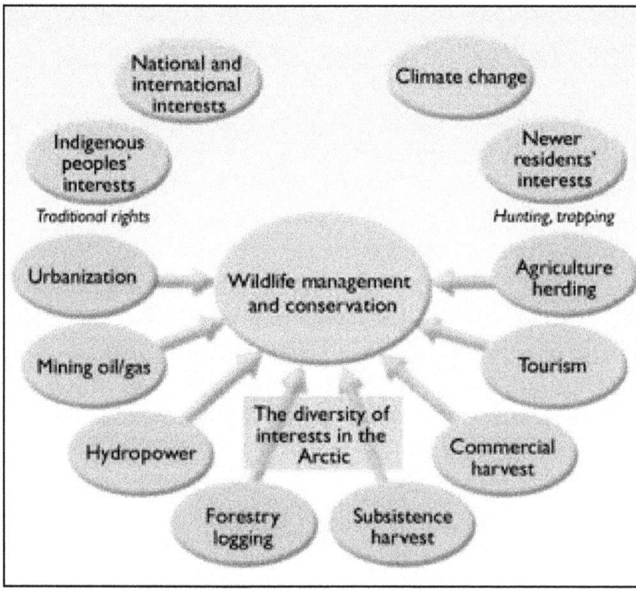

Birds, land mammals, amphibians, reptilies etc, make up the second major group. Migratory bird conservation is managed in an unusual manner, being conducted co-operatively by federal and provincial governments under the authority of the Canada Migratory Birds Convention Act. This law also ensures international coordination. Other land-related wildlife is conserved and managed primarily by provincial laws when wildlife is on provincial land, and by the federal Wildlife Act when wildlife is on federal land.

International Wildlife Protection, Conservation and Management

International conventions (such as treaties) and international institutions also work to protect, conserve or manage wildlife domestically in Canada. In addition to the Migratory Birds Convention, already discussed, the Agreement on the Conservation of Polar Bears (1973), the Convention on International Trade in Endangered Species of Wild Fauna and Flora or CITES (1973), the Ramsar Convention (to protect wetlands and associated habitat; Canada signed on in 1981), the Agreement on the Conservation of the Porcupine Caribou Herd (1987) and the Convention on Biological Diversity (1992) are among the conventions that impose obligations on Canada relating to wildlife protection and management.

Why is Wildlife Conservation Important?

Greg Yarrow of Clemson University defines conservation as the act of consuming natural resources responsibly and carefully so they will be available in the long-term. Yarrow notes that conservation is distinct from preservation in the sense that preservation involves maintaining natural areas for their own sake, which includes protecting them from human use and intervention. Yarrow says that all organisms that are not directly acted upon by humans, such as uncultivated plants and undomesticated animals, could be considered forms of wildlife. Wildlife conservation is considered one of the most pressing environment concern today.

Conservation Impacts Daily Life

There are many reasons – both emotional and practical – why wildlife conservation is important. Conservation can have powerful, direct impacts on the daily lives of people around the world, as well as on plant and animal life and the environment as a whole. our planet with countless wild animals, birds, insects, amphibians,and other creatures. What we do to the earth has an impact on these wild creatures and their survival, as well as on our own. Wildlife conservation is imperative if we are to become stewards of the earth, and to create a truly sustainable future for ourselves and for generations to come. The World Conservation Union estimates that there are **more than 16,000 wildlife species** that are currently threatened with extinction. Only 2–4 per cent of all the species that have ever lived still exist today.

Conclusion

The uses and value of wildlife to society are varied. Wildlife is one part of the equation which, together with vegetation and the nonliving environment, constitutes the "balance of nature," i.e, the set of complex natural processes on which human survival depends. Coastal and inland commercial fishing, based on naturally-reproducing populations, is a significant industry.

Wildlife is a basic component of outdoor recreation and part of the national heritage of all Canadians. Wildlife reflects the condition of the environment and constitutes a "barometer" for measuring environmental change. Throughout Canada, wildlife is the legal property of all Canadians. Its ownership is entrusted in law to the stewardship of the various governments. Sound protection, conservation and management of all wildlife are thus the rightful concern and responsibility of all Canadians.

Ways you can Help and Wildlife Conservation Organizations to Support

1. **World Wildlife Fund** — www.worldwildlife.org
2. **Operation Migration** — This organization helps endangered whooping cranes and Canadian geese by helping them to relearn their migratory routes. The inspiration behind the beautiful and successful movie "Fly Away Home" (available through their website)
3. **Wildlife Conservation Society** — www.wcs.org
4. Explore and enjoy nature and learn more about wildlife at your local nature center, Community Park, or the local chapter of your favorite conservation organization. The **National Audubon Society** puts on a wide variety of nature programmes for all ages, in addition to their conservation activities.
5. **Support your local zoo** – Zoos are important for protecting wildlife and for conducting research to find solutions for the preservation of endangered species.

References

Alexander Burnett (2003). A passion for wildlife: The History of the Canadian Institute of Resources.

Canadian Institute of Resources Law (2006). Canada Wildlife Law Occasional Paper Series.

Groom, M.J, Meffe, G.K. and Carroll, C.R. (2006). Principles of Conservation Biology (3rd edi.). Sinauer Associates, Sunderland, MA. ISBN 0–87893–518–5.

Hunter, M.L.(1996). Fudamentals of Conservation Biology. Blackwell Science Inc, Cambridge, Massachusetts, ISBN 0–86542–371–7.

Janet Foster. (1978, revised 1998). Working for Wildlife: The Beginning of Preservation in Canada.

M. E. Soule and B. A. Wilcox. 1980. Conservation Biology: An Evolutionary–Ecological Perspective. Sinauer Associatess. Sunderland, Massachusetts.

M.E. Soule. (1986). What is conservation Bilogy? *Bioscience*, 35(11): 727–734.

Potter, Dale R, Kathryn M. Sharpe, John C. Hendee (1973). Human Behavior Aspects Of Fish And Wildlife Conservation– An Annotated Bibliography. U. S. Dept. of Agriculture.p.290.

Soule, Michael E. (1986). Conservation Biology: The Science of Scarcity and Diversity. Sinauer Associates. p.584. ISBN 978–08–8789–3795–0.

Van Dyke, Fred (2008). Conservation Biology: Foundations, Concepts, Applications, 2nd edi. Springer Verlag. p.478. ISBN 978–01–4020–6890–4.

Chapter 16

Gyps Vultures and their Conservation

Abhinav Chauhan and Amita Sarkar*

Department of Zoology,
Agra College, Agra, U.P.
****E-mail: amieeta@gmail.com***

ABSTRACT

Vultures are large bodied birds adapted for soaring flight in thermals. They reproduce slowly, reach the breeding age in four to six years and produce first one egg in each breeding season. They are nature's most efficient scavengers; known to occur in cities, towns, villages, forests and open grasslands. They mainly consume carcass flesh. There are total 23 species of vulture. In which 16 are of old world vultures and 7 are new world vulture.Vultures are basically divided into two groups, namely Old world vultures and New world vultures. The former belong to the family acipitridae and are found inhabiting Africa, Asia and Europe. The latter, on the other hand, belong to the family Cathartidae and are mainly seen in the warm and temperate areas of the America. Old World vultures require open spaces and sources of lift to achieve soaring flight. They are therefore found most commonly in either mountains or plains and savannahs in hotter climate. The Gyps vultures are distributed throughout Africa, Europe and Asia, with none above latitude 50 degrees north. There are a total of nine species of vulture found in South Asia. Four of the Gyps vulture species are only found in Asia. These are Oriental white backed vulture (Gyps bengalensis), Long-billed Vulture (Gyps indicus), Slender-billed Vulture Gyps tenuirostris and Himalayan Griffon Vulture (Gyps himalayensis). Since early 1990's, a catastrophic population crash has occurred in three species; the slender billed vulture (Gyps tenuirostris), the Indian or long billed vultures (Gyps Indicus), and the white rumped vulture (Gyps bengalensis). The pace of decline is so great (about 48 per cent per year) that these three species may not survive the next decade. Few factors such as lack of food, poisoning pesticides heavy metal contaminants persecution, viral disease have caused population collapse at local level but none have been shown to cause massive population collapse. Additional research revealed that an anti-inflammatory drug called diclofenac is the cause for their extinction. Exposures of vultures to diclofenac arises through its veterinary use to treat domestic livestock. A major issue in the phasing out of veterinary use of diclofenac is its use in human medicine as an effective anti-inflammatory and analgesic. An alternative drug, Meloxican, is suggested. This paper deals with the distribution of Gyps vultures and their conservation methods.

Keywords: *Vultures, Gyps, Diclofenac, Meloxican.*

Introduction

Vultures are large bodied birds adapted for soaring flight in thermals. To take flight from ground, vultures has to run first *Gyps* species are long-lived. The maximum recorded life span of Eurasian Griffon (*Gyps fulvus*) in captivity is 37 yrs (Newton, 1979). They reproduce slowly, reach the breeding age in four to six years and produce first one egg in each breeding season. They are nature's most efficient scavengers. They are known to occur in cities, towns, villages, forests and open grasslands. Vultures seldom attack healthy animals. They mainly consume carcass flesh. Particular characteristics of many vultures is a bold head, devoid of normal feathers. This helps to keep the head clean when feeding. Though a repulsive creature at close quarters, vultures glide effortlessly in the sky in the very embodiment of graceful motion (Salim Ali, 1996). There are total 23 species of vulture. In which 16 are of old world vultures and 7 are new world vulture.

About Vulture

Vulture is the name of a bird that mainly preys on dead animals and occasionally hunts its own quarry. It is found across the word, with the exception of Antarctica and Oceania. Vultures are basically divided into two groups, namely Old world vultures and New world vultures. The former belong to the family acipitridae and are found inhabiting Africa, Asia and Europe. The latter, on the other hand, belong to the family Cathartidae and are mainly seen in the warm and temperate areas of the Americans. (Mundy *et al.*, 1992).

Kingdom:	Animalia
Phylum:	Chordata
Class:	Aves
Order:	Falconiformes
Size:	60 to 140 cm (24 to 55 inches)
Wingspan:	Up to 3 m (10 feet)
Weight:	Up to 12 kg (25 lb.)
Diet:	Animal carcasses
Natural Habitat:	Throughout the world, except the continents of Antarctica and Oceania
Age:	15 to 20 years
Age of Maturity:	5 to 7 years
Number of Eggs:	One

List of Old World and New World Vulture Species

Old World vultures

- ✰ Eurasian Black Vulture or Monk Vulture, *Aegypius monachus*
- ✰ Lammergeier or Bearded Vulture, *Gypaetus barbatus*
- ✰ Palm-nut vulture, *Gypohierax angolensi*
- ✰ Griffon vulture, *Gyps fulvus*
- ✰ Indian White-Rumped vulture, *Gyps bengalensis*
- ✰ Rüppell's vulture or Rüppell's griffon vulture, *Gyps rueppelli*
- ✰ Long-billed Vulture or Indian Vulture, *Gyps indicus*
- ✰ Slender-billed Vulture, *Gyps tenuirostris*
- ✰ Himalayan Griffon Vulture, *Gyps himalayensis*
- ✰ White-backed Vulture, *Gyps africanus*
- ✰ Cape Griffon, *Gyps coprotheres*
- ✰ Hooded Vulture, *Necrosyrtes monachus*
- ✰ Egyptian Vulture, *Neophron percnopterus*
- ✰ Red-headed Vulture, *Sarcogyps calvus*
- ✰ Lappet-faced Vulture, *Torgos tracheliotus*
- ✰ White-headed Vulture, *Trigonoceps occipitalis*

New World Vultures, including Condors

- ✰ Turkey Vulture, *Cathartes aura*
- ✰ Greater Yellow-headed Vulture, *Cathartes melambrotus*
- ✰ Lesser Yellow-headed Vulture, *Cathartes burrovianus*
- ✰ King Vulture, *Sarcorhamphus papa*
- ✰ American Black Vulture, *Coragyps atratus*
- ✰ California Condor, *Gymnogyps californianus*
- ✰ Andean Condor, *Vultur gryphus*

Distribution

In the wild the Old World vultures require open spaces and sources of lift to achieve soaring flight. They are therefore found most commonly in either mountains or plains and savannahs in hotter climate.

The *Gyps* vultures are distributed throughout Africa, Europe and Asia, with none above latitude 50 degrees north. *Gyps* vultures were typically widespread and abundant, accounting for the majority of individual vulture sightings in both Africa (c. 90 per cent) and Asia (c. 99 per cent) (Houston, 1985). There are a total of nine species of vulture found in South Asia. Four of the *Gyps* vulture species are only found in Asia. These are Oriental white backed vulture (*Gyps bengalensis*), Long-billed Vulture (*Gyps indicus*), Slender-billed Vulture *Gyps tenuirostris* and

Himalayan Griffon Vulture (*Gyps himalayensis*). Eurasian Griffon Vulture (*Gyps fulvus*) breeds in Eurasia but migrates to Africa and South Asia. The geographic ranges of these vultures overlap (Pain *et al.*, 2003). How ever, Chhangani 2004 have recorded, seven species in and around jodhpur in western Rajasthan, India, viz. The King Vulture (*Sarcogyps calvus*), Cinereous Vulture (*Aegypius monachus*), Egyptian Griffon (*Neophron percnopterus*), Eurasian Griffon (*Gyps fulvus*), Himalayan Griffon (*Gyps himalayensis*), Long-billed Vulture (*Gyps indicus*) and White-rumped Vulture (*Gyps bengalensis*). Until recently their huge abundance in India could be explained by the enormous availability of food in the form of domestic cattle and buffalo carcasses that are, for religious reasons, usually not consumed as meat. In some ecosystems *Gyps* vultures feed predominantly on the carcasses of wild rather than domestic ungulates. For example, in the Serengeti, Tanzania, high population densities of *Gyps* are present and consume more than a quarter of the available ungulate carcasses (Houston, 1985).

Habit and Habitat and Distribution

Vultures perform a vital role in nature's sanitation processes by tearing meat from carcasses before it rots. While feeding themselves they prevent the spread of disease which would affect other mammals, including humans survival and a low juvenile survival (Amadon, 1964). They are also highly sensitive to environmental changes (Houston, 1985).

The Oriental White-backed Vulture (*Gyps bengalensis*) prefers cultivated tracts with scattered houses and refuse tips offer more opportunity for obtaining food (Roberts, 1991). Houston (1985) described *Gyps species* as being responsible for a greater consumption of meat than any of the mammalian carnivores.

The Cinereous Vulture (*Aegypius monachus*) breeds very sparsely throughout Balochistan and tribal areas of North-West Frontier Province. There is some migration of birds to lower Sindh in winter and to the Thar and Cholistan deserts and the Potohar plateau around Rawalpindi. In the late 1950s, 30 Cinereous Vulture (*Aegyplus monachus*) could be counted on Karachi municipal refuse dumps in Ghizri creek area. 15 to 20 birds could be seen in summer around Ouetta rubbish dumps in early 1970s. They leave their wintering grounds by mid-February and returns to their nesting haunts on the higher junipers studded crags on Pakistan western borders (Roberts, 1991). 52 eggs of Cinereous vulture, (*Aegypius monachus*) from various in Balochistan had been taken; all were at elevation of 3,048 meters or higher and collected between 10 March and 27 March. 38 were on cliffs and 14 on the crown of Juniper Juniperus macropoda trees. 34 of these eggs came from the Koh-i-murdar Range which is relatively bare of Junipers which might account for the high proportion of nests on the cliff ledges. Most of these were collected during 1930s and early 1940s and about one-third was white all over and unspotted. A fully fledged young bird was taken from a nest on the Khojak ranges near Chaman at the end of May (Roberts, 1991).

The Eurasian Griffon (*Gyps fulvus*) breeds throughout Balochistan and parts of Sindh, Kohistan and the Salt Range. It is comparatively rare in the north parts of Balochistan during mid-winter, in winter time numbers migrates to the Indus plains. They migrate down to the warmer foothills and plains in winter, where they sympatric with the White-backed vultures, (*Gyps bengalensis*). They breed exclusively on mountain cliffs, spending the spring and summer months in the arid steppic mountain regions of Balochistan (Roberts, 1991). Although the literature suggests that vultures have been common birds in South Asia (Roberts, 1991; Gilbert *et al.*, 2002). The drastic decline of three species of *Gyps* vultures (Oriental White-backed, (*Gyps bengalensis*) *Gyps Long-billed* (*Gyps indicus*) and Slender-billed (,*Gyps tenurostris*) in South Asia over the past decade is probably the first of its kind. These three species are now listed as critically endangered (Birdlife International, 2001).

The establishment of 'nursery' areas, where juveniles congregate prior to breeding, is a not uncommon feature (Mundy *et al.*, 1992). Foraging flights for the south Asian vultures may be lower than in other species, possibly due to the high availability of carcases removing the need to travel long distances in search of food. Advances in radio and satellite tracking are providing new information all the time. These management guidelines have been developed for vultures from the genus *Gyps*, and in particular the three critically endangered species of *Gyps* vultures found in South East Asia. Although many aspects will be relevant to other species of vultures, the bulk of this study refers specifically to this genus.

Population Decline

Since early 1990's, a catastrophic population crash has occurred in three species; the slender billed vulture (*Gyps tenuirostris*), the Indian or long billed vultures (*Gyps Indicus*), and the white rumped vulture (*Gyps bengalensis*). One membering in the tens of millions throughout India. These Asian vultures have declined by over 99 percent and currently number less then 10,000 animals. The pace of decline is so great (about 48 percent per year) that these three species may not survive the next decade. Few factors such as lack of food, poisoning pesticides heavy metal contaminants persecution, viral disease have caused population collapse at local level but none have been shown to cause massive population collapse.

The causes for this precipitous decline was not known until 2004. It had been assumed that a vital infection was spreading through earth species, but autopsies of the dead birds revealed the presence of white crystals on several internal organs. These crystals are made upto uric acid, the same chemical responsible for gout in humans. After an extensive investigation of the more common source of vultures mortality such as firearms and lead poisoning. Then it was determined that there was no connection between those vultures deaths from whatever caused the gout like symptoms and vultures deaths from other causes.

Additional research revealed in 2004 that the animals that had gout like symptoms showed high levels of an anti-inflammatory drug called diclofenac in their systems. This drug is the major cause of vultures mass mortality is diligence contamination of live stock carcasses (Oaks *et al.*, 2004; Green *et al.*, 2004; Shultz *et al.*, 2004). However other causes such a habitat destruction, food storage human persecution, poisoning and pesticide use may have caused a gradual decline in vulture population in the region (Bird life international, 2001).

Declaration of Dicolofenac and Vulture Conservation

Recently scientific evidence indicate that diclofenac (a non-steroidal anti-inflammatory drug) is a major cause of vulture decline. Exposures of vultures to diclofenace arises through its veterinary use to treat domestic livestock. Experiment shows that vultures are highly susceptible to diclofenac and are killed by feeding on carcass of an animal soon after it has been treated with normal veterinary dose of drug.

Declines of *Gyps species* in South Asia, Indian populations have declined by 95 per cent within the past decade, a rate of decline unprecedented among common raptors in its rapidity and one the appears to be continuing (Prakash *et al.*, 2003). There are several possible explanation for high mortality and reduced breeding success of *Gyps species* in southern Asia including food storage, persecution, contaminants and infectious disease.

Conservation

Vultures play an important ecological role through rapid consumption of animal carasses. The loss of a major scavenger from the ecosystem will affect the balance between populations of there scavenger species of result in increase putrefying carcasses. The decline in vulture population has resulted in an increase in number of putrefying animal carcasses and changes in the scavenger population in rural as well as urban areas. This has resulted in associated disease risks for wildlife, livestock and humans.

Anthropogenic disturbances leading to destruction of suitable roosting and nesting trees as well as disturbance have a marked impact on vultures and their survival. The general public is ignorant of the ecological importance of vultures. Conservation efforts cannot be successful without the active involvement of local communities.

Alternatives of Diclofenac

Diclfeonac in the a world wide pharmaceutical market: Dicolfenac is a non-steroidal anti-inflammatory drug (NSAID) used for variety of painful and inflammatory conditions. Based on reports received from pharmaceutical industry, the annual total veterinary pharmaceutical market in the country, as per 2003 estimates, is about ₹1000 cr. The non-steroidal anti-inflammatory segments accounts for ₹30 Crores, of which dicolfenac has a market of ₹20 Crores, with an expected annual growth rate of 10 per cent. The number of animals being treated annually with dicolfenac ingestibles alone is projected at about 100 million.

Constraints in Phasing Out Veterinary Use of Diclofenac

Human use of Diclofenac

A major issue in the phasing out of veterinary use of diclofenac is its use in human medicine as an effective anti-inflammatory and analgesic. While there is no reported human medical condition for which diclofenac is the sole, effective anti-inflammatory analgesic, it is nevertheless true that it is an extremely effective drug in human use (Teng, 2005). Also it is possible that a very small number of persons may experience averse effect with all the close substitutes, and thus may only be administered diclofenac. It is thus not feasible to completely phase out human use of diclofenac, in addition to its veterinary use. So long as human use continues, and so long as human diclofenac is a cost effective substitute alternative to veterinary alternatives of diclofenac.

Considering that the balance of societal well-being considerations would require the conservation of vultures through preventing exposure to diclofenac while retaining human use of diclofenac, a way must be found to prevent diversion of human diclofenac to veterinary use. Since veterinary use of drugs are known to be very sensitive to consideration of cost, a feasible approach would be to ensure through appropriate policy instruments, that human diclofenac is too costly for general veterinary use. If there are some extremely poor patients who may only be treated by diclofenac, a way must be found to subsidize diclofenac use in these cases.

Testing Meloxicam as an Alternative Drug for Declofence

Since the three resident *Gyps* species in Asia (*G. bengalensis, G. indicus,* and *G. tenurisotris*) were critically endangered and schedule/species under the wildlife (Protection) Act, 1972 and too rare and too endangered for extensive safety testing.

A phased programme of safety testing for an alternative drug, Meloxican, was observed in vultures at any administered dose. There was no loss of body weight related to treatment, all blood parameters measured remained within their normal ranges, and there was no significant sub-lethal effects. Meloxicam was thus found to be safe at 2.5 × LD50 of diclofenac for a group of test animals. The LD 50 is one way to measure the short-term poisoning potential (acute toxicity) of a material. To increase confidence in safety of Meloxicam a larger sample of vulture was required to be treated with Maloxicam at an estimated maximum level of exposure (MLE). All birds survived at a higher dose of Meloxicam and no loss of body weight was observe over 7 days. All blood parameters measured remained within their normal range (at 2 and 7 days after exposure).

Further trials provides strong evidence that Meloxicam is a safe alternative, better than diclofenac for the treatment of livestock with reduced risk (Swan, 2006). A collaborative project between forest department, Haryana, IVRI and BNHS with technical inputs from RSPB was started in 2005 at the vulture conservation and breeding centre at Pinjore, Haryana for conducting safety trials on Meloxicam (Swaroop *et al.*, 2007).

Conclusion is that oral administration of Meloxicam at a dose equal to or more than the amount may be considered a safe for vultures.

References

Ali, Sálim (1996). The Book of Indian Birds (12th ed.). Bombay: Bombay Natural History Society. ISBN 0–19–563731–3.

Amadon (1964). The incubation period is estimated to be about 50 days. Woodcoci, An assessment of several linear change detection techniques for mapping forest mortality using multitemporal tandsat TMDATA. *Remote Sens. Environ.* 56: 66–67.

Chhangani, A.K. and S.M. Mohnot (2004). Is diclofenac the only cause of vulture decline. Current Science, 87 (11): 1496–1497.

Gilbert, M., Virani, M. Z., Watson, R. T., Oaks, J. L., Benson, P. C., Khan, A. A., Ahmed, S., Chaudhry, J., Arshad, M., Mahmood, S. and Shah, Q. A. (2002). Breeding and mortality of Oriental Whitebacked Vulture Gyps bengalensis in Punjab Province, Pakistan. *Bird Conserv. Int.* 12: 311–326.

Green, R. E., Newton, I., Shultz, S., Cunningham, A. A., Gilbert, M., Pain, D. J. and Prakash, V. (2004). Diclofenac poisoning as a cause of population declines across the Indian subcontinent. *J. Appl. Ecol.* 41: 793–800.

Houston, D. (1985). Indian white–backed vulture Gyps bengalensis. Pp. 465–466 in I. Newton and R. D. Chancellor, eds. Conservation studies on raptors. Cambridge: International Council for Bird Preservation. Technical Publication No. 5.

Mundy, P., Butchart, D., Ledger, J. A. and Piper, S. E. (1992). The vultures of Africa. South Africa: Acorn Books pp. 368–370.

Newton, I. (1979). Population ecology of raptors. Berkhamsted, UK: Poyser.

Oaks J. Lindsay, Gilbert Martin, Virani Munir, Watson Richard, Meteyer Carol, Rideout Bruce, Shivaprasad H.L., Ahmed Shakeel, Chaudhury Mohammad Jamshed Iqbal, Arshad Muhammad, Mahmood Shahid, Ali Ahmad, Khan Aleem Ahmad, (2004). Diclofenac Residues as the cause of vulture population decline in Pakistan, Letters to Nature www.nature.com/nature

Pain, D. J., Cunningham, A. A., Donald, P. F., Duckworth, J. W., Houston, D. C., Katzner, T., Parry–Jones, J., Poole, C., Prakash, V., Round, P. and Timmins, R. (2003). Gyps vulture declines in Asia, temporospatial trends, causes and impacts. *Conserv. Biol.* 17: 661–671.

Prakash, V., Pain, D. J., Cunningham, A. A., Donald, P. F., Prakash, N., Verma, A., Gargi, R., Sivakumar, S. and Rahmani, A. R. (2003). Catastrophic collapse of Indian white–backed *Gyps bengalensis* and longbilled *Gyps indicus* vulture populations. *Biol. Conserv.* 109: 381–390.

Roberts, T. J. (1991). The birds of Pakistan. Volume 1. Non–Passeriformes. Oxford: Oxford University Press Sarrazin, F., Bagnolini, C., Pinna, J. L., Danchin, E. and Clobert, J. (1994). High survival of Griffon vultures (*Gyps fulvus fulvus*) in a reintroduced population. *Auk* 111: 853–862.

Shultz, S., Baral, H. S., Charman, S., Cunningham, A. A., Das, D., Ghalsasi, D. R., Goudar, M. S., Green, R. E., Jones, A., Nighot, P., Pain, D. J. and Prakash, V. (2004). Diclofenac poisoning is widespread in declining vulture populations across the Indian subcontinent. Proc. Roy. Soc. Lond.

Swan, G., Naidoo, V., Cuthbert, R., Green, R. E., Pain, D. J., Swarup, D., Prakash, V., Taggart, M., Bekker, L., Das, D., Diekmann, J., Diekmann, M., Killian, E., Meharg, A., Patra, R. C., Saini, M. and Wolter, K. (2006b). Removing the threat of Diclofenac to Critically Endangered Asian vultures. (2006) *PLOS Biology.* March 2006 4(3): 66.

Swarup, D., Patra, R.C., Prakash V., Cuthbert, R., Das, D., Avari P, Pain, D.J., Green R.E., Sharma, A.K., Saini, M., Das, D., Taggart M. (2007). The safety of meloxicam to critically endangered Gyps vultures and other scavenging birds in India. *Animal Conservation,* 10: 192–198.

Chapter 17

Wild Tigers: A Battle for Survival

Jayita Thakur, Darshan Malik and Shashi Nijhawan

Department of Biochemistry, Shivaji College, University of Delhi, New Delhi

Plans to protect air and water, wilderness and wildlife are in fact plans to protect man

Stewart Udall

ABSTRACT

Tigers are the top predators in the ecosystem which they inhabit and are a precious commodity today, as their population is on a steep decline. Out of original nine subspecies of tigers, three have already become extinct and it is predicted that within the next decade all the tigers may become extinct if the natural habitat is not preserved. The number of tigers survivng currently is placed at around 3,200–a 93 per cent decrease in just a hundred years. While tigers are natives to Southeast Asia and China, India boasts of the highest number of these magnificent animals and this is attributed to the tiger protection initiative called a Project Tiger which was started in India in 1973 to control their diminishing population.

Keywords: *Tigers, Asia, Conservation, Habitat, Illegal trade, Traditional medicine.*

Introduction

The largest of the Asian big cats, the tiger (*Panthera tigris*) is heavily impacted by anthropogenic activities and other natural factors. Wild tigers are mostly localised to Asia, their residence ranging from cold to tropical climates. The beginning of the 20th century boasted of the presence of nine subspecies of these wild cats, namely Balinese, Caspian, Javan, Bengal, Siberian, South-Chinese, Malaysian, Indo-Chinese and Sumatran, consisting of nearly 100,100 members of the species in total. Of these, the Balinese, Caspian and the Javan have become extinct at the turn of the century. Of the remaining sub-species of the wild tigers, almost all of them are endangered in varying degree (Chundawat *et al.*, 2011). With the worldwide tiger population at an all time low, the number of breeding females is estimated to be around 1000. The remaining population is small and fragmented, and extensively subjected to various stochastic events that reduce the survival of breeding adults (Bay *et al.*, 2014; Walston *et al.*, 2010; Sanderson *et al.*, 2006; Kitchener and Dugmore 2000). Apart from the normal human activities, poaching has been a major threat to the entire species, with around 105 tigers being killed each year since 2000 (Verheij *et al.*, 2010).

The tigers primarily use their senses of sight and hearing to stalk and hunt for their prey. Their length may vary from 4–10 feet, and weight ranges between 220–660 pounds. The color of tiger's coat varies between orange/yellow to white. The white coat color is observed in the Bengal tiger subspecies. Tigers are solitary animals, maintaining their territories, which they mark by various methods. They are strong swimmers, unlike others members of the species, and can hunt their prey in water (Shoemaker *et al.*, 1997). The female tigress delivers 2–3 cubs at a time and raises and protects the young by herself (Novak. 1999).

The tigers have played an important role in folklore and mythology of the Asian countries. Their importance is further stressed upon by the fact that the tiger

Balinese – EXTINCT

Caspian – EXTINCT

Javan – EXTINCT

Bengal – Endangered

Siberian (Amur)–Endangered

South China – Endangered

Malaysian – Endangered

Indo-Chinese – Endangered

Sumtran – Critically Endangered

is the national animal of India, Bangladesh, Vietnam, Malaysia and South Korea. The activities like agricultural expansion, timber cutting, building new roads, human settlements, industrial expansion, construction of dams, etc. have severely hindered their territory. Poachers, diseases and climate are some factors which further compound the situation. Being at the top of the food chain, extinction of tigers adversely affect the balance of the entire ecosystem. Hence, it is vital that the tiger, among other wildlife fauna facing extinction, is given special attention as they require to maintain harmony between nature and mankind.

Decline of the Tigers – Hunting as a Sport

The tiger was among the five big game animals that were popularly hunted for sport in the Asian countries. The British and the Indian royalty have been instrumental in decimating the wild tiger population in India – once a stronghold for tigers. It was a matter of pride and a show of valor aristocracy to have killed hundreds of tiger in a lifetime. A trip to India by any British royal family member was considered incomplete without the *shikar* of tigers. It has been recorded that 39 tigers were shot in 11 days in 1911, during the visit of King George V. The Maharaja of Surguja had claims of having recorded to bagging the heads of 1150 tigers in his lifetime (Nowel and Jackson, 1996). Man-eating tigers also posed a threat due to which tigers were killed. Hunters like Jim Corbett had employed hunting to protect villagers from becoming prey to these man eaters (Corbett, 1944).

Decline of the Tigers – Loss of Habitat

Tiger habitat is rapidly declining, getting degraded and fragmented by a range of human activities. The number of tigers that survive in small, scattered islands of habitat, make them ineffective as predators and lead to a higher risk of inbreeding. This reduces genetic diversity and creates a bottleneck situation.

The massive explosion of population is a major source of concern and cause for wildlife habitat destruction. As the rate of deforestation increases to accommodate people, build roads, expand agriculture etc., the wild tigers and communities living near forests get engaged in a conflict. The tigers resort to hunting the domestic livestock, or worse hunt humans. In retaliation, the threatened villagers often poison, shoot or at ensnare the encroaching tigers. The balance between wildlife conservation and community development becomes difficult to attain. Wildlife reserves have been established by the governments of various countries to remedy this situation. However, illegal hunters, known as poachers, pose a serious threat to such an arrangement (http://www.tigersincrisis.com/habitat_loss.htm).

Decline of the Tigers – Poaching

The demand of tiger body parts for traditional Chinese medicine in some countries, and their fur for decorative purpose cause them to be "laundered". Unlike tiger skin, tiger bones are easy to transport – they can be powdered into an odorless form, whereby it is difficult to distinguish. Since killing of wild tigers is cheaper than farming them (Bulte and Damania, 2005) and many Chinese believe that bones from wild tigers offer better effects than farmed tigers (Ellis, 2005), poaching of tigers has become a rampant practice. The rate at which tiger poaching is occurring, the species will become extinct soon unless strong measures are adopted.

The governments are trying to rectify the situation by trying to put a stop to poaching. However, as the market for tiger parts is huge and involves colossal amounts of money, better organization, planning, riddance of corruption is the need of the hour.

Remedies: A Cause for Optimism

The Government of India has taken a pioneering initiative for conserving its national animal, the tiger, by launching the 'Project Tiger' in 1973. From 9 tiger reserves since its formative years, the Project Tiger coverage has increased to 47 at present, spread out in 18 of our tiger range states. This amounts to around 2.08 per cent of the geographical area of our country. The tiger reserves are constituted on a core/buffer strategy. The core areas have the legal status of a national park or a sanctuary, whereas the buffer or peripheral areas are a mix of forest and non-forest

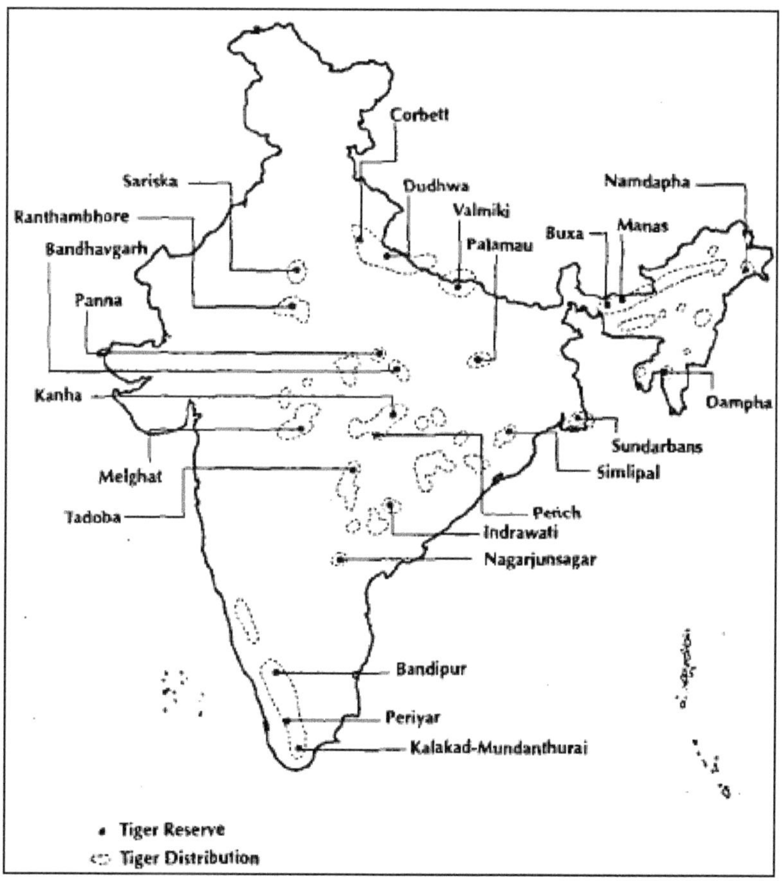

Tiger Reserves in India. Note the unwitting fragmentation of the habitats.
http://projecttiger.nic.in/content/107_1_Background.aspx

land, managed as a multiple use area. Project Tiger aims to foster an exclusive tiger agenda in the core areas of tiger reserves, with an inclusive people oriented agenda in the buffer.

Project Tiger is an ongoing Centrally Sponsored Scheme of the Ministry of Environment, Forests and Climate Change providing central assistance to the tiger States for tiger conservation in designated tiger reserves.The National Tiger Conservation Authority (NTCA) is a statutory body of the Ministry, with an overarching supervisory/coordination role, performing functions as provided in the Wildlife (Protection) Act, 1972.The ADG (Project Tiger) and his officers also service the NTCA. The Regional Offices of the NTCA have been recently established at Bengaluru, Guwahati and Nagpur, each headed by an IGF and assisted by an AIG.

The allocation for Project Tiger during the XII Plan is ₹1245 crore. The expenditure during 2012–13 and 2013–14 are ₹163.87 crore and 169.48 crore respectively.

Due to concerted efforts under Project Tiger, India has the distinction of having the maximum number of **tigers in the world (at present 1706) as per 2010 assessment,** when compared to other tiger range countries. The 2010 country level tiger assessment has also shown a 20 per cent increase of tigers in the country (from 1411 in 2006 to 1706 in 2010). However, there is a decline in tiger occupancy (12.6 per cent) in other areas of tiger States. The tiger corridors for gene flow have been mapped in the GIS domain.

Conclusion

Tigers need jungles, dense forest and thick long foliage to hide and hunt for animals and water to drink. Due to an explosive growth of population in Asia, dense vegetation, jungles and water bodies are fast disappearing and with diminishing wilderness the population of tigers is also declining. Moreover the Tiger reserves are isolated and the chances of inbreeding increases and make the animal more prone to diseases. Poaching and illegal hunting have further led to a decrease in number of tigers. Strict law enforcement to control the illegal international trade of wildlife is required to save the tigers. Awareness among people and more wildlife conservation laws should be made to make Indian wildlife sanctuaries a safe habitat for tigers.

References

Bay RA, Ramakrishnan U, and Hadly EA (2014). A Call for Tiger Management Using "Reserves" of Genetic Diversity. *Journal of Heredity* 2014: 105(3): 295–302 doi: 10.1093/jhered/est086

Bulte EH, Damania R. (2005). An economic assessment of wildlife farming and conservation. *Conservation Biology* 19: 1222–1233.

Chundawat, R.S., Habib, B., Karanth, U., Kawanishi, K., Ahmad Khan, J., Lynam, T., Miquelle, D., Nyhus, P., Sunarto, S., Tilson, R., Wang, S. (2011). "*Panthera tigris*".IUCN Red List of Threatened Species. Version 2011.2. *International Union for Conservation of Nature.*

Ellis R. (2005). Tiger Bone and Rhino Horn: The Destruction of Wildlife for Traditional Chinese Medicine. Washington (DC): Island Press.

http://www.allaboutwildlife.com/endangeredtigersfacts

http://www.tigersincrisis.com/habitat_loss.htm)

Kitchener AC, Dugmore AJ. (2000). Biogeographical change in the tiger, Panthera tigris. Anim Cons. 3: 113–124.

Man–Eaters of Kumaon (1944). Jim Corbett, Oxford University Press, Bombay.

Novak, R. M. (1999). Walker's Mammals of the World. 6th edition. Baltimore: Johns Hopkins University Press. ISBN 0–8018–5789–9.

Nowell K and Kackson P, (1996). Wild Cats, IUCN Gland, Switzerland.

Sanderson E, Forrest J, Loucks C, Ginsberg J, Dinerstein E, Seidensticker JP, Leimgruber MP, Songer M, Heydlauff A, O'Brien T, et al. (2006). Setting priorities for the conservation and recovery of wild tigers: 2005–2015. The technical assessment. Washington (DC): WCS, WWF, Smithsonian, NFWF–STF

Shoemaker, A.H., Maruska, E.J. and R. Rockwell (1997). Minimum Husbandry Guidelines for Mammals: Large Fields. American Association of Zoos and Aquariums

Tiger_Species_WWF Worldwildilife.org

"Tiger". Smithsonian National Zoological Park. Retrieved 7 April 2014.

T.N. Koshoo(1997)Conservation of India endangered mega animals: Tiger and Lion. Current science 73: NO10

Verheij P, Foley KE, Engel K. (2010). Reduced to skin and bones: An analysis of tiger seizures from 11 tiger range countries (2000–2010). Cambridge (UK): TRAFFIC International.

Walston J, Robinson JG, Bennett EL, Breitenmoser U, da Fonseca GAB, et al. (2010) Bringing the Tiger Back from the Brink–The Six Percent Solution. PLoS Biol 8: e1000485. doi: 10.1371/journal.pbio.1000485

Wood, G (1983). The Guinness Book of Animal Facts and Feats. Guinness Superlatives. ISBN 978–0–85112–235–9.

Chapter 18

Fighting for Survival: Greater One-horned Indian Rhinoceros

Renu Baweja, Shashi Nijhawan and Darshan Malik

Department of Biochemistry,
Shivaji College, University of Delhi, New Delhi

ABSTRACT

Rhinoceros, often abbreviated as rhino were once found throughout Euroasia and Africa but their number is fast declining due to illegal wildlife trading of their horns. There are five species of Rhinos and are in danger of becoming extinct because of hunting by the poachers for their horns. The horns are made up of protein Keratin, people grind the horns and the powder is consumed believing that it has therapeutic value. Wildlife trade and loss of habitat is major threat to Rhinos and countries should make sincere efforts to protect their natural resources and curb illegal trading.

Keywords: *Conservation, Diet habitat, Rhinoceros.*

Introduction

Rhinoceros, which were widespread across Asia's tropical forests and Africa's savannas have been highly impacted by illegal International trading of their horns. Today, their population has declined due to Rhino poaching which has increased unprecedently over the past few years and this is attributed to growing demand for horns from Vietnam. Out of five species of rhino: White Black, Sumatran, Javan and Greater one-horned; two Asian species-Javan and Sumatran are critically endangered. Efforts to conserve the Greater one-horned (or Indian) rhino has led to an increase in its number and its status has changed from endangered to vulnerable.

Species of Rhino

Indian Rhinoceros (*Rhinoceros unicornis*) are commonly called as one-horned Rhino since they have one horn rather than two. They are one of the smaller

Black Rhino

Sumataran Rhino

Javan Rhino

White Rhino

Greater One-horned Rhino

Figure 18.1: Species of Rhino.
***Source*: https://www.savetherhino.org/rhino_info/species_of_rhino.**

Rhinoceros species and were found predominantly in parts of Northern India and Nepal. Today, their population has been drastically reduced due to human activities. They are now found in the region of tall grasslands and forest that surround the Himalayas mountain range.

General Characteristics

The Indian Rhinoceros skin is tough with many folds that look like armour. Their skin is brownish-grey in color with wart-like bumps on the neck, shoulders and legs [Figure 18.2]. Its maximum thickness is around 4 cm with their subcutaneous fat 2–5 cm thick and this helps in thermoregulation. The presence of large number of blood vessels under the tissues in folds gives it pinkish color. They have little hair on the hide; few are present on the edges of the ears, the eyelashes, and the tuft on the tail. Males have huge neck folds [1]. Their body length is approximately 12 feet with shoulder height of 6 feet. Their horn length varies from 20–60 cm length with weight of about 3 kg. Their horn is composed primarily of keratin protein and hence very strong. They use their horns for their own defense and also to tear branches during feeding [3]. Horn is not present since birth and it starts to appear about one year after the birth. Their body weight is around 2 tons with the tail of 28 inches. There is a presence of horn on they weigh 2 tons nose also which is short and dull. Their lips are adapted to grasp the grass they eat and they are called prehensile lips [Figure 18.2].

The short and dull horn is also present on the nose. They have poor eyesight with exceptional sense of hearing and smell. This enhanced sense of smell is used to seek out other rhinos during the mating season and also to alert them to the presence of predators. They can move very quickly when aroused and can cover a distance of 48 km/hr. Despite their heavy weight, they are flexible enough to jump or change direction very quickly. Moreover, they are also great swimmers [1].

Location

Historically, Indian Rhinoceros were present across the Northern part of Indian subcontinent. They were predominantly present in the Southern part of Nepal and Bhutan. Due to the change in climate and habitat destruction, their population has gradually been reduced in these regions and by the 19th century, few species survived only in Southern Nepal, northern Uttar Pradesh, Bihar and

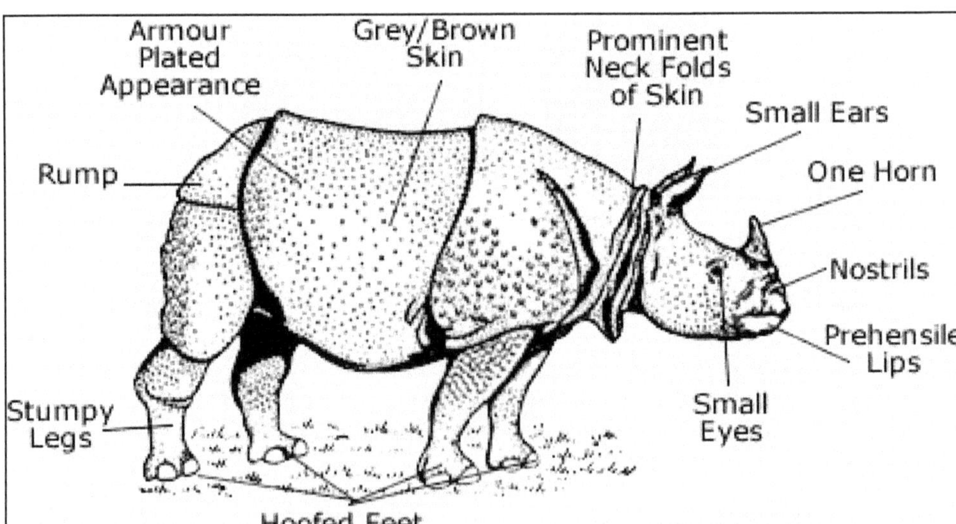

Figure 18.2: Anatomical Features of Indian Rhinoceros.
Source: http://animalcorner.co.uk/animals/indian-rhinoceros/ [6]

Bengal. Some of the species were also present in the Brahmaputra Valley of Assam. In the present scenario, their range has further reduced to few pockets of Northern Bengal, Southern Nepal and Brahmaputra Valley [1].

Habitat

They are semi-aquatic. They reside on grassland in forests, swamps and riversides, after sometimes the region where they live overlaps with other animals. The size of this region depends on the quality of their habitat and ranges from 2–8 sq km. They immerse their body in water or mud holes to avoid heat and eat during cooler part of the day [2]

Diet

The Indian Rhinoceros is herbivore. They mainly feed on grasses, leaves and branches. Their prehencile upper lips are used to they feed mainly on curl around grass stem to bring them into mouth [7]. They also eat fruit and farm crops. Since some of them live near water, they consume omi aquatic plants also. They drink water daily and are fond of mineral licks [8]

Social Behavior and Territory

Except for some females with their young ones, they tend to be solitary. Males territory is not well define however they live alone and defend it aggressively. Many times they fight defined instad of define. put a full stop after defined. and then However, omit it after defend for their territories which often lead to death of one of the rhinoceros. Sometimes these territories may overlap with others. Depending upon the climate or season, these territories change [8]. Females can also move in and out of these territories. Their dung heaps serves as communication points and mark territorial boundaries. Since several animals defecate at the same spots, such a dung heap can become up to five meters wide and one meter high. These Rhinoceros then scratch their hind feet in the dung and through their walk; they carry their smell all around the paths [8]. They also produce some characteristic voice in the form of snorts, roars and honks.

Indian Rhinoceros Reproduction

Males and females are sexually mature at the age of 9 years and 4 years respectively. There is no specific season for breeding. The whistle made by female is an indication for male that she is ready to mate. The gestation period is 15–16 months and a single calf is born at the intervals of about three years. At birth, calf weighs 58–70 kg. The calf remains with its mother for about 2 years before becoming independent. Mother rhinos are very protective and attentive and travel with their young ones; however, males usually travel alone. Their average life span is 35-40 years [6].

Economic Importance and Threat

Rhinoceros horns are used as a medicine against fever and pain. They are also used as aphrodisiac. Their hide, blood, urine and dung also have economic value. There are three major threats:

1. **Hunting:** Hunting was one of the major factor for heir decline. They were being hunted for sport by both Europians and Asians. They were also killed as agricultural pests in tea plantation [4]
2. **Poaching**: Poaching due to the economic importance of their horn is also a reason for declining their population
3. **Habitat loss and destruction**: there are many reasons for the destruction and loss of their habitat [5]:
   (*a*) Invasion by plants into grasslands
   (*b*) Reduction in the extent of grasslands and wetlands habitats due to woodland encroachment.
   (*c*) Grazing by domestic livestock

Conservation Status

Historically, Indian Rhinoceros were found across the entire northern part of the Indian sub-continent. However various human activities severely reduce their population. This pushed the species very close to extinction in the early 20th century and by 1975 there were only 600 individuals surviving in the wild [10]. Today, the Indian rhinoceros is considered to be an endangered animal. Due to rigorous conservation efforts taken by Indian and Nepalese governments along with World Wide Fund for Nature (WWF) and other non-governmental organizations, their number started increasing since 1975. By 2012 their population has been increased to over 3,000 across various regions of India and Nepal. Eighty five per cent of their population lives in India with around 75 per cent lives in the state of Assam. Some of the protected areas are: Kaziranga National Park, Pobitora Wildlife Sanctuary, Orang National Park and Manas National Park. It also occurs in Dudwha National Park in the Indian state of Uttar Pradesh, and in the Jaldapara Wildlife Sanctuary and Gorumara National Park in the state of West Bengal. In Nepal, approximately more than 500 greater one-horned rhinos occur in Chitwan National Park, Bardiya National Park and the Suklaphanta Wildlife Sanctuary [12]. However, poaching and habitat loss and degradation remain a major threat.

With the support of the International Union for the conservation of the nature–status survey and conservation (IUCN-SSC), Asian Rhino Specialist Group, an Indian Rhino Vision (IRV) 2020 and a Nepal Rhino Action Plan have been developed. These cover a number of important and specific conservation measures, including: translocating rhinos to bolster struggling populations (*e.g.*, Manas National Park) and to start new populations; improving security around rhino populations and reducing poaching; assessing habitat status and management needs; expanding available habitat through active management; improving protected area infrastructure; training staff in specific rhino conservation techniques; reducing human-wildlife conflicts; involving local people in rhino conservation; and implementing education and awareness programmes [12]. Apart from this there are many organizations, foundation or programme working in the protection and conservation of Rhinoceros including CITES [13], International Rhino foundation [14], Save the rhino [15], WWF [16], etc.

Proposed Conservation

It is important to maintaining the wild population. There is need for continuous monitoring their numbers. Though many organizations/projects are working for their protection, they will only be successful if reasons for their downfall will be eliminated. Efforts to control poaching, improve habitat will be helpful for their conservation. There is immediate need to run educational programmes specially to teach local communities how to co-exist with the rhinos and use the land sustainably. Moreover research to find an alternative to horn used as medicine will also be an additional help to reduce the poaching and hence further increase their population.

Conclusion

Illegal poaching of rhino is driven by high prices paid for their horn. People in west believe in a myth that rhino horn act as aphrodisiac and sexual stimulant. The horns have been used in traditional Chinese medicines for over 2000 years but sudden spike in demand for horns is due to belief of Vietnamese that rhino horn can cure cancer and terminal illness. To address the challenge of reducing number and protection of the rhinos, the Government should increase the law enforcement and impose strict deterrent and awareness should be created among people about the myths.

References

1. Foose, T. J. F., Van Strien, N. (eds.). 1997. Asian Rhino Status Survey and Conservation Action Plan. IUCN SSC Asian Rhino Specialist Group. IUCN, Cambridge UK.
2. http://animalcorner.co.uk/animals/indian-rhinoceros/
3. http://animaldiversity.org/accounts/Rhinoceros_unicornis/
4. http://a-z-animals.com/animals/indian-rhinoceros/
5. http://en.wikipedia.org/wiki/Indian_rhinoceros
6. http://wwf.panda.org/what_we_do/endangered_species/rhinoceros/asian_rhinos/indian_rhinoceros/
7. http://www.animalfactguide.com/animal-facts/indian-rhinoceros/
8. http://www.edgeofexistence.org/mammals/species_info.php?id=11
9. http://www.edgeofexistence.org/mammals/species_info.php?id=74
10. http://www.indiantravelportal.com/wildlife/indian-rhinoceros.html
11. http://www.iucnredlist.org/details/19496/0
12. http://www.rhinos.org/indian-rhino-vision-2020
13. http://www.savetherhino.org/rhino_info/species_of_rhino/greater_one-horned_rhinos/factfile_greater_one-horned_rhino

14. http://www.worldwildlife.org/species/greater-one-horned-rhino
15. International Rhino Foundation. (July 2005).
16. Save the Rhino. (July 2005).
17. WWF. (July 2005).

Chapter 19

Pangolin: A Unique Mammal with Scales

Deepika Yadav[1], Darshan Malik[2] and Sunita Singh[2]

[1]Department of Zoology, [2]Department of Biochemistry
Shivaji College, University of Delhi, New Delhi

"Earth provides enough to satisfy every man's needs, but not every man's greed."

— Mahatma Gandhi

ABSTRACT

Surrounded with mysteries due to its secretive, solitary and nocturnal habits, pangolin or scaly anteater is considered as the most illicitly trafficked and traded mammal in today's world despite all the provided legal protections. Pangolins are insectivorous mammals distributed in the Afrotropical (sub-Saharan Africa) and Indomalayan regions, with eight known species.Habitat loss and hunting for illegitimate international trade, largely driven by traditional medicinal market in East Asia are the major causes of their destruction. Due to the unfortunate belief that pangolin's meat, scales and foetuses are luxurious delicacies pertaining to miraculous medicinal values, these magnificent creatures are doomed with extinction.

***Keywords**: Conservation, Mammals, Pangolin, Therapeutic values.*

Introduction

Pangolins or *"the pine cone on legs"* are endangered species that is living on the edge of extinction as the wild stocks of the species are rapidly depleting. It is an inscrutable and highly intelligent animal also known as a "mischievous escape artist" but luck has finally run out as this astonishing creature is being slaughtered on an industrial scale and faces extinction.[16] Occurring in African and Asian continents, pangolins have been exploited for food and traditional medicines through the history and this exploitation continues even today.[22] Illegal trade in South Asia, however, has earned these scaly mammals an unenviable tag of the most trafficked animal on earth [Figure 19.1]. Given their secretive habits, pangolins have been subjected to very little conservation or natural history research so slight is known about their biology, ecology and conservation needs. As a result, pangolin populations in Asia are precipitously declining, and they have been extirpated from vast areas. Inter-continental trade in Africa to Asian markets is an added worry. In response to this predicament, various precautionary measures in conjunction with Wildlife Reserves and locals are being taken to seize the extinction of the species because "**once extinct is gone forever**". The ongoing devastation of the slow and strange pangolin is a grim image of the long-lasting impact of human greed on fellow inhabitants on the planet Earth.

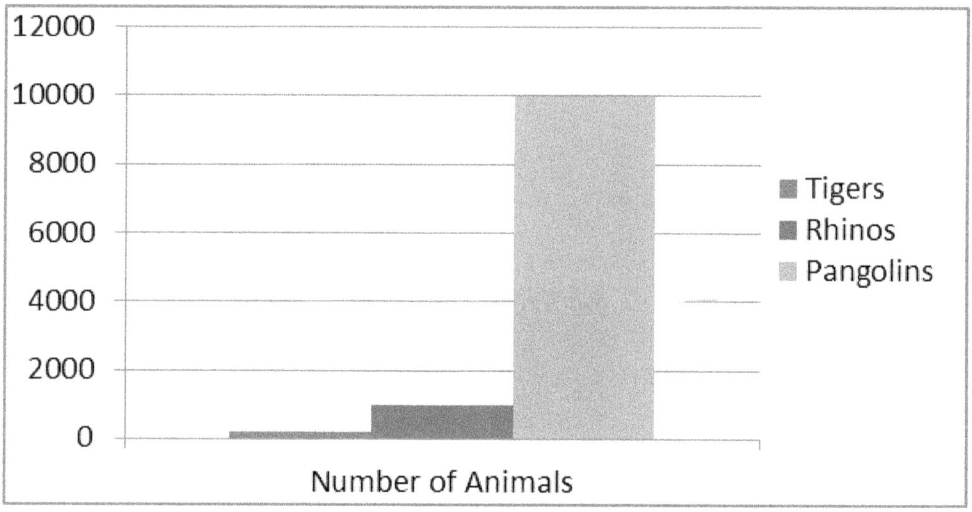

Figure 19.1: Wildlife Trafficking Estimates (per year). Adapted from https://thisbookisunwritten.wordpress.com/ 2015/02/17/the-pangolin- problem-2/.

Nomenclature and Classification

The name, "pangolin", is derived from the Malay word "pengguling", which means "something that rolls up", referring to the animal's natural defense method of rolling into a tight ball when threatened [16]. This practice although protects them from their natural hunters in wild but makes them vulnerable to humans traffickers who simply pick and collect them up in this position.

Worldwide eight species of Pangolin are in existence from one extant family, Manidae while several extinct species contribute to their 80 million-year-old evolutionary pathway [Figure 19.2]. The eight species together comprise one Order: Pholidota [9]

Scientific Classification

Kingdom: Animalia

Phylum: Chordata

Class: Mammalia

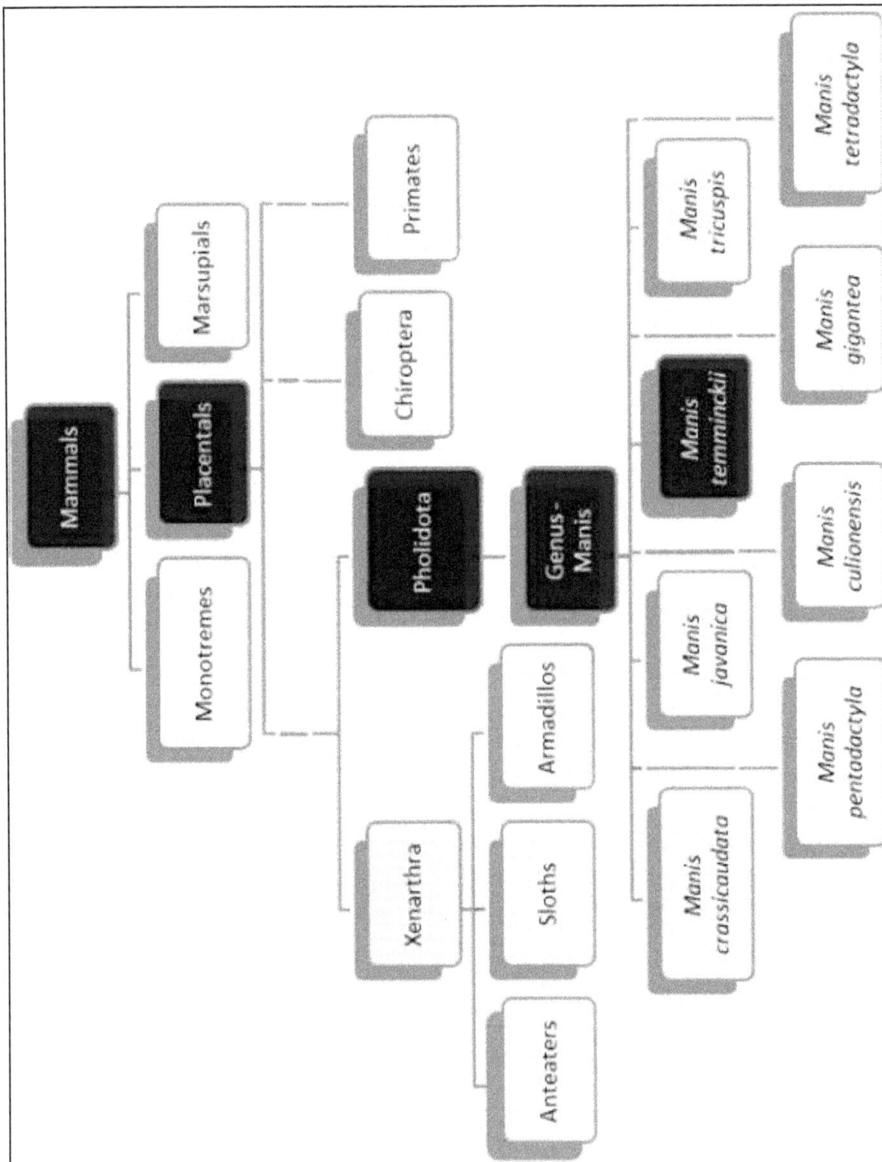

Figure 19.2: Pangolin: Schematic Classification.

***Source:* http://bioweb.uwlax.edu/bio203/s2012/grosshue_crai/Pictures/phylogenetic_tree2.jpg.**

Infraclass: Eutheria

Superorder: Laurasiatheria

Order: Pholidota

Family: Manidae

Genus: Manis

Subgenus Manis:

- ✰ Indian pangolin (*M. crassicaudata*)
- ✰ Chinese pangolin (*M. pentadactyla*)

Subgenus Paramanis:

- ✰ Philippine pangolin (*M. culionensis*)
- ✰ Sunda pangolin (*M. javanica*)

Subgenus Smutsia:

- ✰ Giant pangolin (*M. gigantea*)
- ✰ Ground pangolin (*M. temmincki*)

Subgenus Phataginus:

- ✰ Tree pangolin (*M. tricuspis*)

Subgenus Uromanis:

- ✰ Long-tailed pangolin (*M. tetradactyla*)

Various Species of Pangolin

Showing in Figure 19.3

Conservation Status

With the exception of *M. culionensis*, all 4 Asiatic pangolin species were classified as 'lower risk: near threatened' by the International Union for Conservation of Nature (IUCN 2007) and appeared in Appendix 2 of the Convention on International Trade in Endangered Species of Wild Fauna and Flora, with a zero annual export quota for wild caught individuals or those traded for commercial purposes by the Convention on International Trade in Endangered Species of Wild Fauna and Flora (CITES).[4] But due to the habitat destruction and therapeutic usage of various body parts, the Chinese Pangolin and the Sunda Pangolin are now classified as Critically Endangered on the International Union for Conservation of Nature (IUCN) Red list. It is the worst listing on the Red List when a species approaches extinction. Indian and Philippine pangolins are upgraded to endangered status on the scale.[1] Previously the four African species were also listed as "least concern" or "near threatened" but now are classified as "vulnerable".[1,2,20,21] As Asian pangolin species has become scarce and rare now resulting from illegal hunting and poaching practices, wildlife traffickers are now moving towards the African parts for exploiting African species for fulfilling the needs of illegal markets. It is an alarming situation in both the native continents. No wonder that pangolin

Figure 19.3: The Eight Pangolin Species
***Source*: Novataxa.blogspot.fr**

1: Temminck's ground pangolin *M. temminckii*; 2: Chinese pangolin *M. pentadactyla;* 3: Sunda pangolin *M. javanica;* 4: Philippine pangolin *M. culionensis;* 5: Giant pangolin *S. gigantica;* 6: Indian pangolin *M. crassicaudata;* 7: African black-bellied or Long-tailed pangolin *P. tetradactyla;* 8: African white-bellied or Tree pangolin *P. tricuspis* [24].

is considered as one of the most endangered and most heavily trafficked CITES-protected mammal on Earth.

Existing Species of Pangolins along with their Previous/New Conservation Status

Asian Species

1. Chinese or Formosan pangolin (*Manis pentadactyla*) (EN) (CR)
2. Malayan or Sunda pangolin (*Manis javanica*) (EN) (CR)
3. Indian or thick-tailed pangolin (*Manis crassicaudata*) (NT) (EN)
4. Philippine or Palawan pangolin (*Manis culionensis*) (NT) (EN)

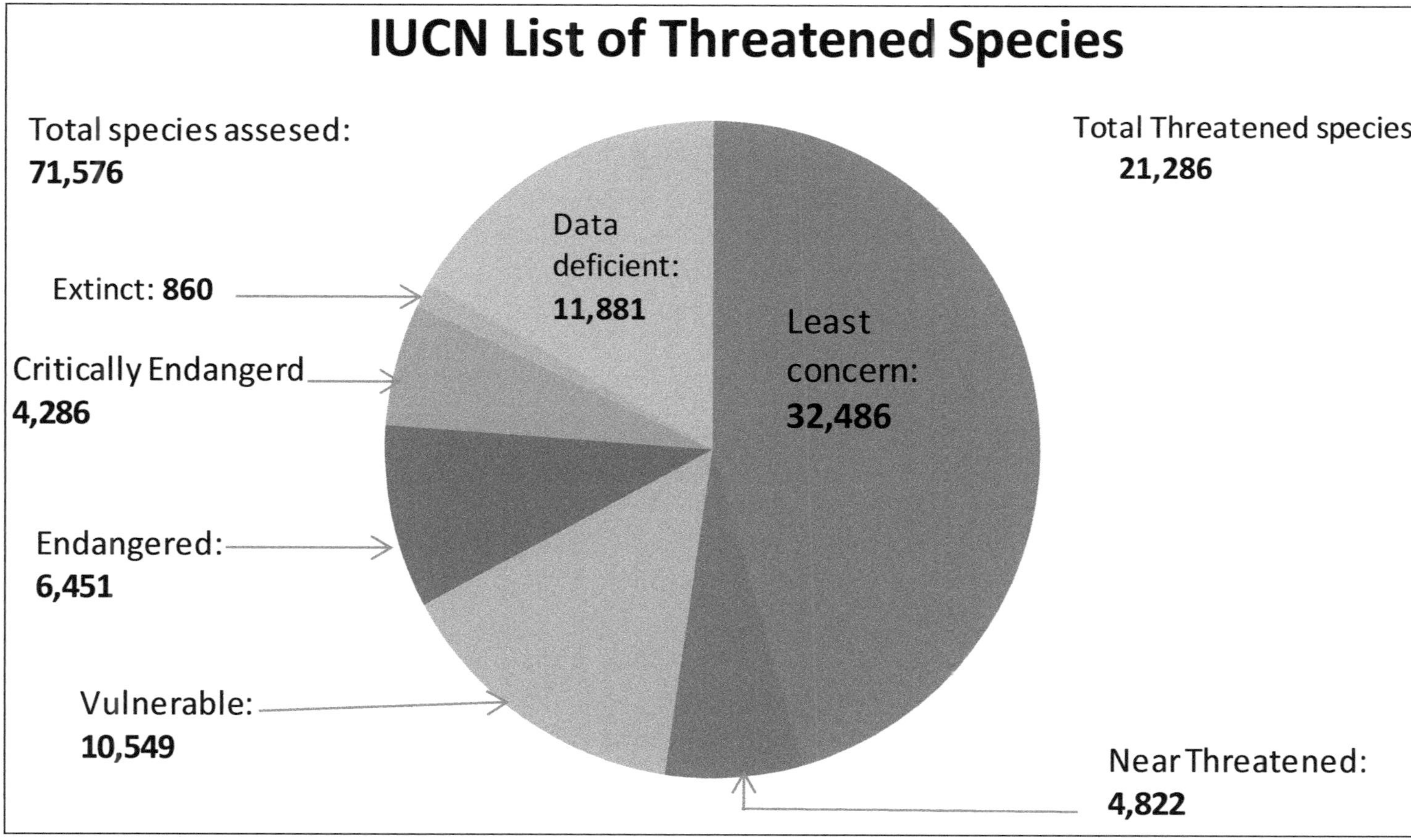

Figure 19.4: The Total Number of Animal Species on Earth is Estimated is 8.7 million. Adapted from: National Geographic/IUCN.

African Species

5. African white-bellied or Tree pangolin (*Phataginus tricuspis*) (NT) (VU)
6. Giant ground pangolin (*Smutsia gigantea*) (NT) (VU)
7. Temminck's ground or Cape pangolin (*Manis temminckii*) (LC) (VU)
8. African black-bellied or Long-tailed pangolin (*Phataginus* tetradactyla*) (LC) (VU)

where,

EN = Endangered; CR = Critically Endangered: NT = Near Threatened; VU = Vulnerable; LC = Least Concern) *Or *Uromanis*[5,7]

Geographical Distribution and Habitat

Four pangolin species occur across Asia: the Indian pangolin, the Chinese or Formosan pangolin, the Malayan or Sunda pangolin and the Palawan pangolin. Whereas four species are found in south of the Sahara Desert in Africa: the Cape or ground pangolin, the tree pangolin, the giant pangolin and the long-tailed pangolin. [9]

Pangolins are found in a vast range of habitats including tropical and flooded forests, thick bushy, clear cultivated areas and savannah grassland; generally they occur where large numbers of ants and termites are found.The eight living species of pangolins include:

- ✰ Indian pangolin (*Manis crassicaudata*): The Indian pangolin inhabits plains, hills, forests and grasslands of India, Pakistan, Shri Lanka and Nepal. The species not only endures at high altitudes of up to 2300 meters but also withstands dry habitats. The Indian pangolin is a solitary and nocturnal terrestrial pangolin species that forages on the ground.
- ✰ Chinese pangolin (*Manis pentadactyla*): The Chinese pangolin is a deep burrowing pangolin that inhabits India, Indochina, Bhutan,Taiwan, China, Bangladesh Nepal and Myanmar. It digs deep burrows in the ground with the help of its strong and clawed fore feet.
- ✰ Sunda pangolin (*Manis javanica*): The arboreal Sunda pangolin inhabits South East Asian region (Indonesia, Thailand, Malaysia, Singapore, the Philippines, Vietnam, Laos and Cambodia).
- ✰ Philippine pangolin (*Manis culionensis*): The Philippine pangolin inhabits the terrestrial regions such as the forests and grasslands of the Philippine Islands.
- ✰ Giant pangolin (*Manis gigantea*): The giant pangolin inhabits equatorial Africa from West Africa to Uganda, including Kenya and Tanzania. It inhabits a vast range of habitats from savannah, rainforests and other forests although it prefers lower elevations.
- ✰ Ground pangolin (*Manis temmincki*): The ground pangolins are inhabitants of Africa. Of the four African pangolins, the ground pangolin is the only one whose range extends into Southern as well as the Eastern parts of the

continent. The ground pangolin is nocturnal and terrestrial. It inhabits savannahs, open habitats and sparse woodlands.

- Tree pangolin (*Manis tricuspis*): The tree pangolin inhabits Guinea, Sierra Leone along with Western and Central Africa. The tree pangolin is an arboreal and nocturnal species that lives in lowland tropical moist forests.
- Long-tailed pangolin (*Manis tetradactyla*): As suggested by the name itself, the long-tailed pangolin has a very long tail yet is paradoxically the

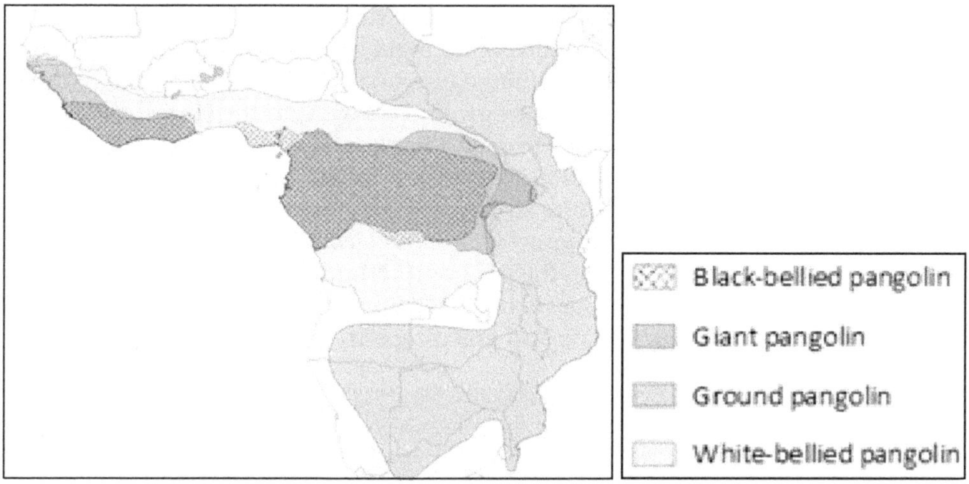

Figure 19.5: Geographical Distribution of African Pangolin.
***Source*: SCALING UP PANGOLIN CONSERVATION; IUCN SSC Pangolin Specialist Group Conservation Action Plan; Daniel W. S. Challender, Carly Waterman and Jonathan E. M. Baillie; July 2014.**

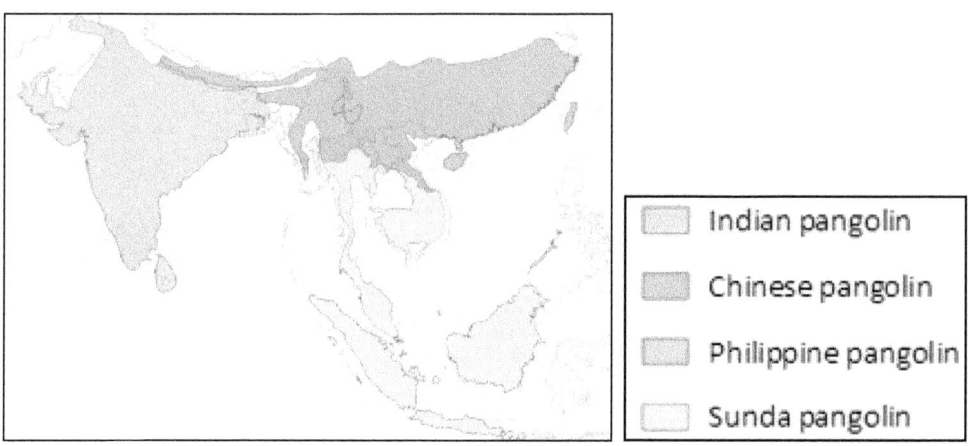

Figure 19.6: Geographical Distribution of Asian Pangolin.
***Source*: SCALING UP PANGOLIN CONSERVATION;IUCN SSC Pangolin Specialist Group Conservation Action Plan;Daniel W. S. Challender, Carly Waterman and Jonathan E. M. Baillie;July 2014.**

smallest of all the eight species of the pangolins. It lives in west central Africa [27].

General Characteristics

Pangolins, often called "scaly anteaters," are solitary, shy, slow-moving, predominantly nocturnal mammals covered in tough and overlapping scales. Of the eight species of pangolin only one, the long-tailed pangolin is regularly active during the day while the rest are nocturnal and have very small eyes relatively to their body size. They locate termites and ants with a strong sense of smell and hearing. Giant ground pangolins are the largest of all eight species. Long-tailed pangolins weigh around 2 to 3 kg. The black-bellied pangolin is the smallest species about 80 cm long, including the tail and the giant ground pangolin is the largest, extending up to 1.8 meters. Found in tropical Africa and Asia, some species dig underground burrows while others live in hollow trees. They forage mostly at night and rest during the day. Pangolins climb with strong, gripping forelimbs and a tail that curls over outstretched branches for support in search of swarms of leaf nesting weaver ants. If approached by any danger, the pangolin rolls into a tight and impenetrable ball with its tail and legs covering its head for thorough protection. They have a long prehensile tail used for seizing and grasping purposes.[24,25,27] [Figure 19.7].

Figure 19.7: Pangolin: A Pine Cone on Legs.
Adapted from: blogs.andbeyond.com/suzie_mackley/the-elusive/-and-extrardinary-pangolin/

Hard scales, tiny ears and thick skin help them to ward off ant bites as it devours an entire nest.

Physical Characteristics

The scaly-plated pangolin resembles a reptile more than a mammal. Along with its armoured shell and peculiar walk, the pangolin looks moreover like a medieval battalion than an animal under threat. All eight pangolin species vary in size from about 1.6kg (~3.5 lbs) to about 33kg (~73 lbs) maximum while color varies from light yellowish brown to olive and dark brown. Overlapping scales are made up of keratin which cover and protects most of the body of the animal. Resembling Artichoke leaves, the scales keep on growing throughout the life like hairs do. The edges of scales are regularly filed down while digging out burrows and tunnel in search of food through sand. Pangolin undersides are devoid of scales and are covered with thin and sparse fur. Asian pangolins have thick bristles emerging from in between the scales while their African counterpart lacks them. Females are generally smaller than the males and have one pair of mammae.

Pangolins are insectivorous, predating on ants and termites. They exhibit a number of anatomical adaptions to this specialized diet which includes a conical shaped head, strong claws for breaking apart ant nests and termite mounds, small eyes with thick eyelids and a remarkably long, muscular, sticky and rod-shaped tongue ideal for reaching, scooping, lapping or flicking out ants and termites in deep crevices. Pangolins have poor vision, so they locate termite and ant nests with their strong olfactory senses. They have special voluntary muscles that can seal their nostrils, ears and mouth to protect them from ant or termite attacks. Pangolin's tongue is attached near its pelvis and last pair of ribs and is longer (up to 1½-3 feet long) than the whole body when fully stretched. The tongue retracts into a sheath in its chest cavity when not in use. A pangolin's stomach is muscular with keratinous spines (a compensation for teeth in the animal) and small stones, which mashes and

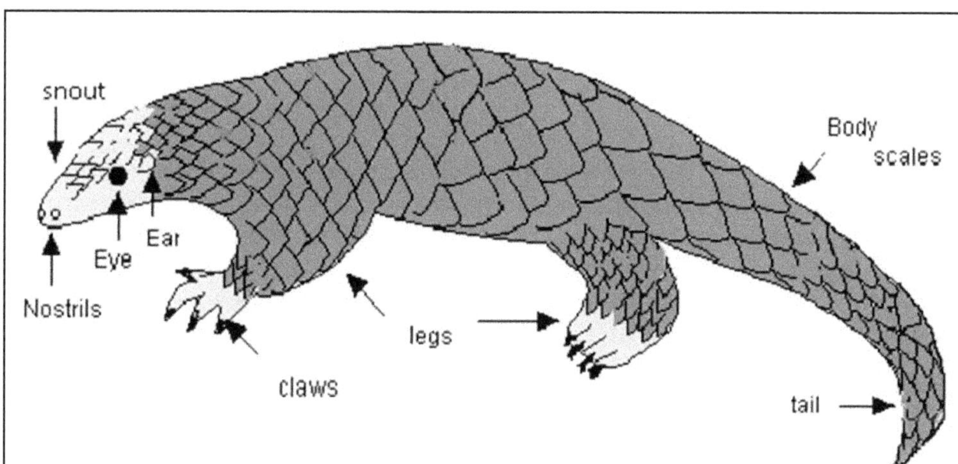

Figure 19.8: Pangolin: General Anatomy.
© Deepika Yadav.

grinds prey in the same manner as a bird's gizzard. Pangolin limbs are stout and adapted for digging. Each paw has toes with long, curved claws used to dig the nests of termites and ants and to dig nesting and sleeping burrows for itself. They are fast runners and accomplished swimmers, while some species are completely terrestrial like the African ground pangolin, others, such as the African tree pangolin, are proficient climbers (arboreal), using their claws and semi-prehensile tails to grip bark and scale trees [9,10,11,25] [Figure 19.8].

Ecology

Pangolins are among the most valuable and widely traded taxa in the Southeast Asian illegitimate wildlife trade, yet little is known of their ecology and they are rarely reported in biodiversity surveys. Eight different pangolin species are found across Asia and sub-Saharan Africa.

Ecological Significance

A. Control of the Termite and Ant Population, Lesser Insecticide Pollution

Pangolins' insatiable appetite for insects renders them a significant role in their ecosystem: Pest Control. One adult pangolin can consume more than 70 million insects per year. These insects cause some drastic damage to human society and for curbing this, insecticide is applied. Pangolin, therefore, plays in important role in ecosystems by regulating social insect populations. So pangolin provides an environment friendly natural mechanism to check these insect menaces.

B. Increasing Biodiversity

Burrowing animals are ecologically considered as significant species globally, as their actions create breeding habitat or shelter for many other animals and are thought to contribute to increased species diversity. Pangolin's digging for feeding process also helps to loosen and aerate the soil which results in more vegetation propagation. Their action improves the nutrient quality of the soil and also aids the decomposition cycle [30,36].

Diet

It is an insectivore that principally survives on ants and termites, which they may supplement with various other invertebrates including bee larvae, flies, worms, earthworms and crickets. This specialist diet makes them extremely difficult to maintain in captivity as they often reject unfamiliar insect species or become ill when fed distant food. They also have special muscles in their mouths which prevent ants and termites from escaping after captured by the animal [10].

Life-span

It is not known that how long pangolins can live in the wild because of their solitary, secretive and nocturnal habits, though it is presumed that the pangolins lives as long as twenty years in wild since in captivity they are able to live for about

19 years maximum. They are under stress and depression in captivities like zoos and are tend to be malnourished as well because of the peculiarity of their diet [10].

Social Behaviour

Pangolins are not often observed in the wild due to their solitary, secretive and nocturnal nature [11]. It usually sleeps during the day in hollow trees or burrows, and emerges in evening to forage. Pangolins dig big, deep and circular chambered burrows for sleeping and nesting. Malayan pangolin also sleeps in the hollows and forks of trees and logs. Their winter burrows are strategically excavated near termite nests that provide a lasting food source during the harsh weather. They emit loud hissing and puffing sound when frightened or angry. When any danger approaches, they roll up into balls to protect themselves, lashing out with their tail, the scales on which can easily cut open a predator's skin. They are also capable of emitting a noxious musky-smelling yellow acid from anal glands. Pangolins identify their territories by scent marking with urine, secretions from anal glands and by scattering feces. These odours may display dominance and sexual status and also help individuals in recognizing each other [10,25,27].

Figure 19.9: Pangolin in a Defensive Position.
***Source*: Project Pangolin.**

Reproduction

As foresaid, the pangolins are mammals, so they give birth to the young ones. Male and female pangolins are sexually dimorphic which means the sexes differ. In most species, males are 10-50 per cent heavier than females, while Indian pangolins can be up to 90 per cent heavier. During the mating period, females and males may share the same burrow and show some diurnal activities exceptionally. Scaly anteaters attain sexual maturity after two years, and typically give birth to a single offspring per pregnancy. Their gestation period lasts between 65 -150days, varying by the species and the placenta type is the diffused one.

Figure 19.10: A Pangolin Mother Carrying the Baby.
***Source*: https://www.flickr.com/photos/floridagirl7/3737188885.**

Young are generally weaned at around three months of age.Usually newborn babies weigh around 85–450 g (3–18 ounces) at birth and measure roughly 30 cms, depending on the species [16]. The young ones are born with the open eyes at the time of birth and have soft scales that eventually harden rapidly. The young ones could not walk with legs; rather they move with the help of abdomen. During the age of 6-8 weeks the young ones often eject a pungent yellow secretion from their anal glands that is used to restrict predators and other animals from taking advantage of their mothers. The mother pangolin carries her young on her tail [Figure 19.10].

Scales, their Chemical Composition

The **scaly anteater** is an exclusive mammal possessing scales. These unique scales are α and β keratinized complex structured horny derivative of the epidermis [Figure 19.11]. It is divisible into three separate regions. The dorsal plate forms approximately one-sixth part of the scale total thickness. Scales are made up of flattened solid keratinized cells without basophilic nuclear remnants which make

Figure 19.11: Scales of Pangolin.
***Source*: www.pangolinsg.org.**

the region skirmish. The dorsal plate contains bound phospholipids and sulfhydryl groups but has weak disulphide bonds. The major part of the scale is composed of the intermediate plate formed of less flattened cells without basophilic nuclei. This region is particularly rich in disulphide bonds having fewer amounts of bound phospholipids or sulfhydryl groups. The ventral plate is of few cells thickness with bound phospholipids opulence also occurring in the underlying scale bed epidermis. These three distinct regions of the scale are formed from separate epidermal germinal areas which are not developed into a granular layer. Keratohyalin granules are formed in the epidermis in between the scales. The pangolin scales are probably homologous with primate nails based upon the histological structure and chemical composition. Pangolin scales are particularly effective as body armour, as they can curl into a ball for defence and the scales have sharp edges in addition to being tough [12,13]. The most noticeable characteristic of the pangolin is its colossal, scaled armour which covers its whole upper body with the exception of the belly and the inside of the legs. These rigid protective scales, made of keratin keep on growing throughout life. It has up to 160–200 scales, about 45 per cent of which are placed on the tail. Scales can be 6.5–7 cm long, 8.5 cm wide, and weigh 7–10 grams. The skin and scales make up about one-fourth to one-third of the total body mass of the pangolins.[12,13] The pangolin scales consist of stearic acid, cholesterol, N-butyl tricosylamide, cyclo (L-seryl-L-tyrosyl), cyclo (D-seryl-L-tyrosyl), and other 18 kinds of microelements like zinc, sodium, titanium, calcium, lead, silicon, phosphorus, iron, manganese, chromium, magnesium, nickel, copper, barium, boron, aluminum, molybdenum, and tin.

Keratin is the same material of which human fingernails, hairs, rhino horns and bird talons are made up of and account for 20 per cent of pangolins weight. The scales are very hard and protect pangolin against animal predators. Pangolin sales account for 20 per cent of the entire wildlife black market.

Table 19.1: Amino Acid present in Pangolin Scales

Name of Amino Acid	*Pangolin Scales Contents (wt per cent)*
Glycine	10.6–13.7
Alanine	2.3–3.0
Valine	3.2–4.0
Leucine	7.8–7.9
Isoleucine	2.0–2.3
Serine	5.0–5.5
Threonine	3.2–3.4
Tyrosine	17.2–20.0
Phenylalanine	3.0–3.2
Cysteine	1.9
Methionine	0.4–0.6
Tryptophan	0.4–0.5
Arginine	8.0–8.6
Histidine	0.7–1.2
Lysine	2.5–2.6
Aspartic acid	7.1–7.9
Glutamic acid	10.4–10.9
Proline	8.6–9.7

Source: http://link.springer.com/article/10.1007/BF00462216#page-1.

Threats

Pangolins are given the unenviable tag of being the most illegally-traded mammal in the world. According to the International Union for the Conservation of Nature (IUCN), virtually no information is available on population levels of any species of pangolin due to their secretive, solitary, and nocturnal habits and there has been a meagre research on their population densities. However, all species are thought to be in decline, with some more rapidly than others, particularly the Asian species.They has a low fecundity rate of only one offspring per year so the global pangolin population is diminishing at faster rate than that it could be recovered. Apart from the destruction and deterioration of natural habitat by rapid urbanization, pangolins are being eaten to extinction by East-Asian diners. The demand for pangolins comes most conspicuously from China and Vietnam, where pangolin scales are unfortunately believed to be a cure-all of sorts and the animal flesh is eaten as a luxurious culinary delicacy, but a greater risk arises from the belief that the scales and various body parts have certain therapeutic values. Scaly anteaters are heavily poached and are in peril because of the devastating illegal international trade. Various parts of the pangolin are valued sources of food and medicine. The scales are used as an aphrodisiac or made into rings or charms.[11]] There is a huge demand for pangolin scales as nearly all of their body parts are used in food, medicine, and manufacturing leather goods. Malayan pangolins are the most numerous mammals found in confiscations of illicit wildlife shipments in Southeast Asia. Over one million individuals are estimated to have been poached over the past one decade which makes pangolins the most illegally traded mammal in the world. In spite of a commercial trade ban for wild-caught pangolins in Asia, the illegal trade is rampantly thriving. Consumers are ever willing to pay increasingly high prices for their meat. The illegitimate trade in pangolin species has reached up to an epic scale, with the Chinese and Sunda pangolins now classified as Critically Endangered. As the populations of the four Asian pangolin species plunge, traders are now turning to Africa for meeting the growing demand. Conservationists are mounting concern that African pangolins will increasingly be targeted to meet the supply of the ever expanding far East Asian demand, as Asian species populations continue to plummet towards extinction. Conservationists warn there is already evidence of an underground, intercontinental trade of pangolins between Africa and Asia. African species are largely hunted as bush meat plus some sort of regional use of their scales and other body parts in folk medicines and cultural traditions and rituals is there. The flesh of both adult and fetuses is erroneously believed to be blessed with health benefits. It is hard to find how many pangolins are left in the wild in the world as every year more and more are ending up in cooking pots.

Supposed Traditional Therapeutic Values

Pangolin various body parts are used to treat many ailments in traditional therapeutics [35] [Figure 19.12]. Amongst these are excessive nervousness (depression), Kleptomania, hysterical crying in children, women possessed by evil spirits, malarial fever, relieving palsy, ejection of placenta, menstrual disturbances, food poisoning, epilepsy, asthma, deafness and draining pus [6,8]. Their scales, blood and other body parts are also widely used in Traditional Chinese Medicines (TCMs), considered as aphrodisiac and 'health tonics' [14, 26, 34, 36]. Despite the proven scientific fact that keratinous body parts of other animals (*e.g.*, rhino horns) are void of any medicinal or curative properties, pangolin scales are continuously consumed as a remedy for a wide variety of health problems such as reducing swelling, improving blood circulation, rheumatism and stimulating lactation in breast-feeding women [32]. Pangolin-derived TCMs are also rumoured to have cancer curing properties. The dearth of endangered species has formed a "bottleneck" for traditional Chinese medicine production. Pangolins along with leopard bone, antelope horns, bears and rare snakes are essential raw ingredients for the treatment of "coronary heart disease, myocardial infarction, angina, stroke, coma, cerebral vascular disease, fever-induced shock, hallucination, infectious diseases and cancer treatment [15, 17].

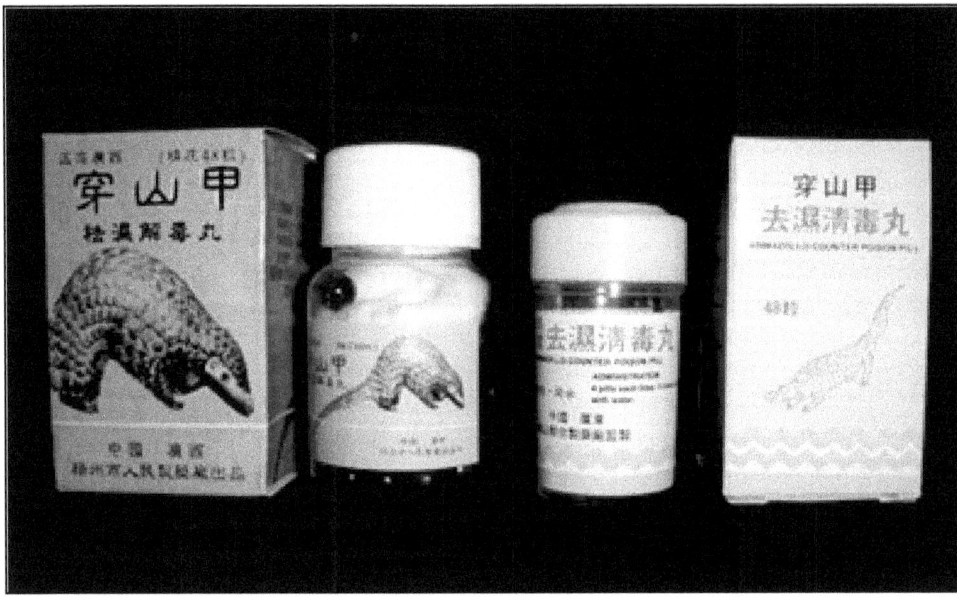

Figure 19.12: Pangolin: Therapeutic Uses in Traditional Medicines.
***Source*: http://www.bushwarriors.org/pangolins-the-other-bushmeat/**

Contradictions of Traditional Remedies

Pangolin scales contain cholesterol, stearic acid, and fatty acid amide compounds so improper use of them can cause liver damage along with the side effects including bloating, loss of appetite, jaundice and some allergy symptoms such as rashes with swollen face.They all possess approximately the same medicinal value that is of a placebo. Harvesting these animals from the wild until they are all gone is a shameful task. Keratin scales has no effect beyond that of a placebo and the only consequence of consuming it is contributing to the eradication of this protected and endangered species. [13]

Conservation: Strategic Challenges and Proposed Solutions

Pangolins are unique in being the world's only scaly mammals, but this uniqueness has dragged them in dire need of urgent conservation action plans and execution as well, at earliest possible manner. Initiatives to address the issue are still in their infancy as is proved by the diminishing numbers of the species. Some well-trained pangolin specialist groups are needed to be a global voice for pangolins by working towards advance knowledge and understanding of pangolins worldwide, their conservation, natural history and ecology. International Union for the Conservation of Nature (IUCN), Species Survival Commission (SSC), the Zoological Society of London (ZSL), San Antonio Zoo, Houston Zoo, TRAFFIC, Pangolin Specialist Group, International Criminal Police Organization (ICPO)-Interpol, World Customs Organization (WCO), the ASEAN Wildlife Enforcement Network (ASEAN-WEN) and Ocean Park Conservation Foundation Hong Kong are among the major groups working towards the direction.[33] World Pangolin

Day is celebrated on 21st February each year since 2012 for raising awareness about these unique mammals and to mark their plight [18,20,30]. Some of the major measures taken for pangolin conservation includes the designing of community awareness programmes aiming the locals, develop their interest in pangolin survival, strengthen and empower law enforcement officers responsible for seizing banned shipments of wildlife, proper training programmes for rangers to properly handle and release confiscated pangolins. While aiding the conservation of pangolins through a wider programme of site protection and broad biodiversity surveys and evolving programmes with a strong focus on raising community awareness, the major obstacles to combat the underground wildlife trade of Asian pangolins include lack of resources and capacity to implement conservation programmes and low prioritization by governments. Communities and local officials need to be fully aware of the benefits of pangolin conservation so that they can be involved in conservation action enthusiastically.

Rehabilitation centers lack the capacity and infrastructure to care for injured and stressed pangolins that are confiscated from the illicit wildlife trade and full proof techniques to care are still being developed. Although conservation efforts emphasize preventing the removal of pangolins from the wild, but wildlife authorities are constantly seizing live pangolins from illegal traders, so it is crucial to develop ways to treat and rehabilitate pangolins in captivity so that they can be safely returned to their native habitats. The conservation efforts should not only focus on preventing poaching and illegal trade but also consider that pangolins rarely survive in captivity so it is crucial to expedite the development and implementation of targeted *in situ* pangolin conservation activities and protecting the habitat of these remarkable animals. There is no evidence supporting the claims regarding medicinal properties of pangolin scales or any other body parts. Outlining or charting strategic actions plans which are critical to the conservation of pangolins and requires urgent implementation including reducing demand for pangolins among consumers, the strengthening of site protection at pangolin strongholds, helping communities move away from poaching pangolins and the reinforcement of legislation. Since chemical composition of pangolin scales is almost similar to the primate nails they can be well substituted as the traditional therapy ingredients or be replaced by some similar properties phytochemicals in the quest to save this marvelous creature.

References

1. All 8 species of pangolins being eaten to extinction, *Posted on:* July 29th, 2014 *by* Save Pangolins, SavePangolin.org
2. Baillie, J., Challender, D., Kaspal, P., Khatiwada, A., Mohapatra, R. and Nash, H. (2014). "Manis crassicaudata". *IUCN Red List of Threatened Species. Version 2014.2*. International Union for Conservation of Nature. Retrieved 2014-07-29.
3. Blogs.andbeyond.com/suzie_mackley/the-elusive/-and-extrardinary-pangolin/
4. CITES 2010. Appendices I, II, and III valid from 24 June 2010. www.cites.org/

5. Corbet, G. B., and J. E. Hill. 1992. Mammals of the Indomalayan region. A systematic review. Oxford University Press, Oxford, 488 pp.
6. Chinese Medicine and the Pangolin _ Abstract _ Nature http://www.nature.com/nature/journal/v141/n3558/abs/141072b0.html
7. Gaubert, P. and Antunes, A. 2005. Assessing the taxonomic status of the Palawan pangolin *Manis culionensis* (Pholidota) using discrete morphological characters. *Journal of Mammalogy* 86: 1068–1074.
8. (*Hong Kong Naturalist,* July 1937, 79) Chinese Medicine and the Pangolin _ Abstract _ Nature
9. http://animaldiversity.org/accounts/Manidae/classification/
10. http://en.wikipedia.org/wiki/Pangolin
11. http://en.wikipedia.org/wiki/Indian_pangolin
12. http://link.springer.com/article/10.1007/BF00462216#page-1
13. http://link.springer.com/article/10.1007/BF00462216#page-2
14. http://onlinelibrary.wiley.com/doi/10.1111/zoj.1967.46.issue-310/issuetoc (Journal of the Linnean Society of London, Zoology, Volume 46, Issue 310, Article first published online: 15 MAY 2008
15. http://pangolins.org/2012/01/16/medicinal-use-pangolin-farms-in-china-photos-video/
16. http://www.disclose.tv/forum/meet-the-pangolin-the-mammal-with-reptile-scales-t84652.html#ixzz3bBE8FxIw
17. http://www.chineseherbshealing.com/pangolin-chuan-shan-jia/
18. IUCN SSC Pangolin Specialist Group website (www.pangolinsg.org)
19. IUCN Pangolin Specialist Group (1996). *Manis javanica*. In: IUCN 2006.
20. IUCN (International Union for Conservation of Nature) (2007) IUCN Red List of Threatened Species. Available at: www.redlist.org (accessed July 2007)
21. IUCN Red List of Threatened Species., from www.iucnredlist.org. Accessed on16 March 2008. http://www.iucnredlist.org/details/full/136497/0
22. Pangolins being eaten to extinction, conservationists warn Environment The Guardian.
23. "Pangolins in peril: using local hunters' knowledge to conserve elusive species in Vietnam"JournalofEndangeredSpeciesResearch., http://www.intres.com/abstracts/esr/v6/n1/
24. Pangolin/8 Pangolin Species - PROJECT PANGOLIN.html
25. PROJECT PANGOLIN: SavePangolin.org
26. Pangolin (Chuan Shan Jia).htm.
27. Pangolins - Manis - The Animal Encyclopedia.

28. Schoppe, S. and R. Cruz. 2008. The Palawan Pangolin *Manis culionensis*,. p. 176–188. In Pantel, S. and C. S. Yun. (Ed.). Proceedings of the Workshop on Trade and Conservation of Pangolins Native to South and Southeast Asia, 30 June–2 July 2008, Singapore Zoo, Singapore. TRAFFIC Southeast Asia, Petaling Jaya, Selangor, Malaysia,. 227 pp.

29. Schoppe, S. and Cruz, R. 2009. The Palawan Pangolin Manis culionensis. *In Pantel, S. and Yun, C.S. (ed). Proceedings of the Workshop on Trade and Conservation of Pangolins Native to South and Southeast Asia, 30 June - 2 July 2008, Singapore Zoo, Singapore. TRAFFIC Southeast Asia, Petaling Jaya, Selangor, Malaysia.*

30. SCALING UP PANGOLIN CONSERVATION, IUCN SSC Pangolin Specialist Group Conservation Action Plan, Daniel W. S. Challender, Carly Waterman and Jonathan E. M. Baillie, July 2014.

31. Schlitter, D. A. (2005). Wilson, D. E., Reeder, D. M, eds. *Mammal Species of the World* (3rd ed.). Johns Hopkins University Press. p. 531. ISBN 978-0-8018-8221-0. OCLC 62265494

32. Soewu and Adekanola, Traditional-Medical Knowledge and Perception of Pangolins (Manis sps) among the Awori People, Southwestern NigeriaJournal of Ethnobiology and Ethnomedicine 2011, 7: 25 http://www.ethnobiomed.com/content/7/1/25

33. TRAFFIC. 2008. Final Call for Pangolins. http://www.traffic.org/home/2008/6/29/final-call-for pangolins.html

34. #Vietnam champagnewhisky.htm

35 WorldHealthOrganisation: TraditionalMedicine.2003[http://www.worldhealthorganisation.com].

36. www.pangolinsg.org http://savepangolins.org/what-is-a-pangolin/# Distribution and Habitat Pangolins and Chinese medicine _ ONCA.

Chapter 20

Wildlife Conservation

Amita Saxena and Smita Chauhan

Fishery Resource Management, College of Fisheries, GBPUAT, Pantnagar – 263 145, Uttarakhand

ABSTRACT

India has a very diverse flora and fauna. It has almost all types of habitats. But due to man's greed and advancement of technology the forests have been depleted this is the reason of endangerment of the species. The purpose of wildlife conservation is to protect wild flora and fauna against the encroachment of expanding human activity.

Keywords: *Biodiversity, Conservation, Lion, Threats.*

Introduction

Wildlife resources constitute a vital link in the survival of the human species and have been a subject of much fascination, interest, and research all over the world. Today, when wildlife habitats are under severe pressure and a large number of species of wild fauna have become endangered, the effective conservation of wild animals is of great significance. Because every one of us depends on plants and animals for all vital components of our welfare, it is more than a matter of convenience that they continue to exist; it is a matter of life and death. Being living units of the ecosystem, plants and animals contribute to human welfare by providing:

1. Material benefit to human life;
2. Knowledge about genetic resources and their preservation; and
3. Significant contributions to the enjoyment of life (*e.g.*, recreation).

Human society depends on genetic resources for virtually all of its food; nearly half of its medicines; much of its clothing; in some regions, all of its fuel and building materials; and part of its mental and spiritual welfare. Considering the way we are galloping ahead, obvious of what legacy we plan to leave for future generations, the future does not seem too bright. Statisticians have projected that by 2020, the human population will have increased by more than half, and the arable fertile land and tropical forests will be less than half of what they are today. Genetic resources are treated as inexhaustible mineral resources, but we need to care about them. It is here that the concept of management and conservation of wildlife comes into play, because anything that is not human or undomesticated is 'wildlife'. Presence or absence of an animal or plant in a region is determined by ecological and historical factors. Animals and plants are living indicators of the characteristics of their environment; their ranges mark the places where environmental conditions are the same or similar. To interpret the range of a species properly, it is necessary to know, in detail, the conditions required for the species to live and thrive. The science of zoogeography has both ecological and historical aspects. On this basis, the world can be divided into six zoogeographical regions

Nearctic: North America and Greenland

Pala arctic: Eurasia, without India

Ethiopian: Africa, south of the Sahara

Oriental: India and Indochina

Australian: Australia and New Zealand

Neo tropical: South and Central America and the Antilles

Wildlife conservation is the practice of protecting endangered plant and wild animal species and their habitats. Among the goals of wildlife conservation are to ensure that nature will be around for future generations to enjoy and to recognize the importance of wildlife and wilderness lands to humans. Many nations have government agencies dedicated to wildlife conservation, which help to implement policies designed to protect wildlife. Numerous independent non-profit organizations also promote various wildlife conservation causes.

WILDLIFE CONSERVATION DEFINED Wildlife protection act started in 1972 Wildlife conservation is the protection of species and habitats of animals. The banning of hunting seasons for endangered or threatened species.

From polar bears and grey whales to buffalo and migratory birds, extraordinary species of wildlife are being threatened by industrialization, overhunting, or reckless population management practices

Concept of wildlife conservation has been around since ancient times. Restrictions on taking game are mentioned in the Bible, first official hunting season may have been established in the 13th century by **Kubla Kahn.**

Today, wildlife conservation has evolved into a science, but its goal remains essentially the same: to ensure the wise use and management of renewable resources. Given the right circumstances, living organisms that we call renewable resources can replenish themselves indefinitely.

Conservation: *Is the wise use of natural resources, without wasting them.*

Importance of Wildlife Conservation

Beauty, Economic value: Timber, fur, tusk, ivory, leather, honey, etc. Scientific value: Gene pool for the scientists to carry breeding programmers in agriculture, animal husbandry and fisheries, also maintenance of ecological balance for eco-tourism.

Biodiversity of India

Biodiversity of India or Wildlife of India is a complex mixture of species of diversified origins. India is home to a number of rare and threatened animal species. Home to about 7.6 per cent of all mammalian, 12.6 per cent of avian, 6.2 per cent of reptilian, and 6.0 per cent of flowering plant species. Tropical rainforest of the Andaman Islands, Western Ghats, Northeast India, coniferous forest of the Himalayan ranges and Dry deciduous forest of central and southern India.

People of the Indian subcontinent were once blessed with some of the most profuse natural gifts: verdant forests, water-stocked Himalayan ranges, rich coastal fish resources, productive estuaries, grassy pastures, and bountiful river systems. Abundant rain and fertile soils added to this plentitude. Years of mismanagement, however, have degraded our forests, wounded our coastline, and poisoned our aquifers with devastating results. Today, India contains 172 Species (2.9 per cent of the world's total number) of animals that are considered to be globally threatened by the IUCN. These include 53 species of mammals, 69 species of birds, 23 species of reptiles, and 3 species of amphibian.

Wildlife Conservation in India

India is the seventh largest country in the world and Asia's second largest Nation with an area purposes. India is divided into 29 states and union territories and is home to more than 1 billion people, which is approximately16 per cent of the world's population. Ecologically, India can be divided:

1. The Himalayan Mountain system;
2. The peninsular India sub-region (woodlands and desert); and
3. The tropical rain forest region.

A great wealth of biological diversity exists in these regions and in India's wetlands and marine areas, this richness is shown in absolute numbers of species and the proportion of the world's total biodiversity.

Group	*Number of Species in India (SI)*	*Number of Species in the World (SW)*	*SI/SW Per cent*
Mammals	350	4629	7.6
Birds	1224	9702	12.6
Reptiles	408	6550	6.2
Amphibians	197	4522	4.4
Fishes	2546	21,730	11.7

List of Animals which are Endangered in India and Uttarakhand

- ☆ Red Panda
- ☆ Asiatic Lion
- ☆ Nilgiri Langur
- ☆ Snow Leopard
- ☆ Himalayan Wolf and Lion-tailed Macaque
- ☆ Hoolock Gibbons
- ☆ Tibetion Antelope
- ☆ The Blackbuck
- ☆ Spot-billed Pelican and Sarus Crane
- ☆ Lesser Flamingo and Great Indian Bustard
- ☆ Indian Elephant and Asian Black Bear
- ☆ Tiger and Indian Rhinoceros
- ☆ Dhole and Capped Langur
- ☆ Blue Whale and Ganges River Dolphin
- ☆ Markhor and Namdapha flying squirrel
- ☆ Wild water buffalo and Servant Mouse
- ☆ Malabar Tree Toad
- ☆ Nicobar Treeshrew and Narcondam Hornbill
- ☆ Gharial Green turtle
- ☆ Gaur and White-bellied Musk Deer
- ☆ Leatherback turtle and Mugger crocodile
- ☆ Clouded Leopard and Eld's Deer
- ☆ Kondana Soft-furred Rat and Salim Ali's fruit bat
- ☆ Malabar Large-spotted Civet

Asiatic Lion

Introduction

The Asiatic lion (*Panthera leo persica*), also known as the Indian lion, is a lion subspecies that exists as a single population in India's Gujarat State. It is listed as *Endangered* by IUCN due its small population size. The lion population has steadily increased in the Gir Forest National Park, more than doubling from a low of 180 individuals in 1974 to 411 individuals consisting of 97 adult males, 162 adult females, 75 sub-adults, and 77 cubs as of April 2010.

The Asiatic lion was first described by the Austrian zoologist Johann N. Meyer under the trinomen *Felis leo persicus*. It is also known as the Persian lion.

The Asiatic lion is one of five big cat species found in India, apart from Bengal tiger, Indian leopard, snow leopard and clouded leopard. It formerly occurred in

Persia, Mesopotamia, Baluchistan, from Sind in the West to Bengal in the East, and from Rampur and Rohilkund in the North to Nerbudda in the South. It differs from the African lion by less inflated auditory bullae, a larger tail tuft and a less developed mane.

The Asiatic lion also known as babbar sher in Hindi. The only place in the wild where this species is found is in the Gir Forest in Kathiawar of Gujarat, India. The Asiatic lion is one of the five major big cats found in India, the others being the Bengal tiger, the Indian leopard, the snow leopard and leopard. They prey mainly on cattle, deer, pig and other herbivore by making a short, high-speed charge up to 80 km per hour.

The Asiatic lions once ranged from the Mediterranean to the Northeastern parts of the Indian subcontinent, but excessive hunting, habitat destruction, decline in natural prey and human interference have reduced their number. This forest is the last refuge of the Asiatic lion in the world, other than those protected in various zoos.

The Asiatic Lion grows to a height of approximately 90 cm, with their length ranging from 200–280cm. Their long and hard tails grow to an additional length of 60 - 90cm. Lions can weigh in at between 200–275 kg. The lions of Gir have made it a habit of killing livestock and sometimes even camels.

Their breeding season is timed along with the onset of winter in the months of October and November. Their gestation period is 3 and a half months, with litters ranging between 3 – 4 in number. The lionesses breed on an average of every two years and are ready to have their first lot of babies around the age of 30–36 months. years They have a life expectancy ranging between 20–30 years Lioness are the main hunters in any pride of lions

The Asiatic lion, in India, is now found in only one pocket located in the Gir National Park of the state of Gujarat. There have been efforts made to relocate a small number of these lions to another park to try and ensure better survival chances of this majestic animal. However, these plans are yet to be put into action.

The Asiatic lion prefers grassy plains, savanna, and open woodlands as its habitat. Like African lions, they are very social and live in groups called "prides." Asiatic lion prides are smaller than African lion prides, and males prefer to associate with the pride when mating or during hunting. They are nocturnal creatures, and most of the day is spent resting. Asiatic lions prey on medium to large mammals, such as wildebeest, zebra, giraffes and gazelles, and young elephants, rhinos, and hippos. Breeding can occur year-round and females give birth to one to five cubs after a gestation period of 100 to 119 days.

Just like their African cousins, the Asiatic males too have handsome manes, which are absent in the females. The Asiatic lions tend to have smaller manes than their African cousins. The Asiatic lions, however, have a bushier overall coat and also have longer tufts of hair at the end of the tail and on the elbow joints. The colouration of the manes vary from lion to lion. The only rare ones are the very dark manes, which according to a research done in Africa are the ones prefered by lionesses!

Asiatic lions have smaller manes than their African cousins.

Lions mostly live in large prides. These prides can sometimes have upto 3 adult males but it is always one that is the dominant leader. The males are known for their laziness and lordship like behaviour. The females do all the hunting, with the males only rarely joining in when the prey is a very large animal like an aggressive buffalo. However, once the kill is made, the males always get the first go at the meal.

The bonding among a family of lions is extremely strong with aunts and sisters helping in the bringing up of all young. Females stay with the pride all through life, whereas the males tend to set out on their own around the age of three. They mostly lead solitary lives then onwards but have been known to also roam territories in pairs and trios. These bachelors are known to be the main threats to the dominant males leading their prides. They are also known to kill cubs to try and get the females into estrus once again. These bachelors are mostly brothers that left a pride together but individual bachelors have been known to team up with other individuals.

Lions, unlike the tiger, hunt in groups. They collectively stalk their prey and have been commonly seen applying strategies that would do any army commander proud. Very often some of the females pinpoint a particular individual prey and chase it in the direction of other lionesses waiting in ambush. The prey is mostly killed by a quick, powerful bite to the spine or with the help of a classic choke grip, with the strong jaws of the lion cutting off air supply to the lungs.

Although history shows the coexistence of lions and tigers, there is no prevalent example of this anywhere in the world at present. Lions do coexist even in the current era with leopards and cheetahs. However, they are extremely territorial and attempt to kill these leopards and cheetahs whenever their paths happen to cross. If the attempt at relocating lions to other parts of India is finally undertaken, it will also answer the question of whether it is possible for two such ferociously territorial and powerful large cats to inhabit the same jungle.

The Asiatic Lion has been declared the most endangered large cat species in the world. Their numbers ranging between 250–300, all concentrated in the same area, they are under the constant threat of being wiped out by some deadly epidemic. It is hoped by all conservationists that the governing authorities settle their differences of opinion on the best possible plan and take some action before it's too late to save one of the must magnificent beasts to roam the planet

Location

The Asian lion is found only in India's Gujarat Province today, predominantly in the Gir Forest National Park and Wildlife Sanctuary. Three protected areas (Gir Sanctaury, Gir National Park and Pania Sanctuary) represent the core habitat of the Asian Lion with an area of 1452 km^2 of forest.

Two further sanctuaries (Mitiyala and Girnar) provide protection for dispersal or satellite areas. A new sanctuary, in the Barda forest is being established in order to provide alternative habitat for Gir lions. These protected areas consists of dry teak forest, deciduous forest, thorny forest and savanna with diverse flora and fauna.

The Asiatic Lion exists as a single isolated population in India's Gujurat State, numbering approximately 175 mature individuals, all occurring within one subpopulation (but in four separate areas, three of which are outside of the Gir Forest protected area). Since the population now extends beyond the boundary of the lion sanctuary, and the numbers are stable, the subspecies is listed as Endangered based simply on the population size (none of the other criteria are met). The subpopulation was in fact increasing for a time but is now considered stable (IUCN Cats Red List Workshop 2007). Constant monitoring is required to ensure poaching levels do not increase; 34 animals were reported killed in 2007 (Jackson 2008).

Threats

Asian lions were hunted, shot, and poisoned to protect domestic animals from possible predation, and this remains a risk to this day. Some poaching occurs and there is evidence of organized gangs forming with the express purpose of poaching these lions.

Asian lions are also in danger from electrified fences that farmers put up to keep grazing animals out of their fields. A large numbers of these magnificent felines have also fallen into the open well shafts dug in the area, where they drown. This is a considerable risk since there are approximately 20,000 wells in close proximity to the Gir Forest National Park.

The Asiatic lion currently exists as a single subpopulation, and is thus vulnerable to extinction from unpredictable events, such as an epidemic or large forest fire. There are indications of poaching incidents in recent years.

There are reports that organized gangs have switched attention from tigers to these lions. There have also been a number of drowning incidents after lions fell into wells.

Prior to the resettlement of Maldhari, the Gir forest was heavily degraded and used by livestock, which competed with and restricted the population sizes of native ungulates. Various studies reveal tremendous habitat recovery and increases in wild ungulate populations following the Maldhari resettlement during the last four decades. Farmers on the periphery of the Gir Forest frequently use crude and illegal electrical fences by powering them with high voltage overhead power lines. These are usually intended to protect their crops from but lions and other wildlife are also killed. Nearly 20,000 open wells dug by farmers in the area for irrigation have also acted as traps, which led to many lions drowning. To counteract the problem, suggestions for walls around the wells, as well as the use of "drilled tube wells" have been made

The range of the lion in North-Africa and South-West Asia formerly stretched across the coastal forests of Northern Africa and from Northern Greece across South-West Asia to Eastern India. Today the only living representatives of the lions once found throughout much of South-West Asia occur in India's Gir Forest (Nowell and Jackson 1996) but there are now also some groups outside Gir Forest like Girnar, coastal subpopulation and Bali Tana subpopulation.

Current total population size is about 350 animals has been increased, but is now stable as it has reached its expansion limits and there are now increasing poaching incidents. At least 100 animals are outside the Gir Forest protected area and total number of mature animals is about 175.

Asiatic lions are genetically distinct from the lions of sub-Saharan Africa, although the difference is not large, being smaller than the genetic distance between human racial groups (Nowell and Jackson 1996). The Gir is dry deciduous forest dominated by teak, the predominance of which is partially due to the silvicultural practices of the Gujarat State Forest Department, which permits logging and replants clear-cut areas with teak (Nowell and Jackson 1996). The forest, which covered about 2,600 km^2 at the turn of the century, has since shrunk to less than half this size. Most of the remaining forest is included in the Gir National Park and Wildlife Sanctuary (Nowell and Jackson 1996).

The Asiatic lion currently exists as a single subpopulation, and is thus vulnerable to extinction from unpredictable events, such as an epidemic or large forest fire.

There are indications of poaching incidents in recent years (there are reports that organized gangs have switched attention from tigers to these lions). There have also been a number of drowning incidents after lions fell into wells.

One of the major threats that the Indian lions are facing is the hostility that suffer at the hands of humans living near the sanctuary.

The lions face the usual threats of poaching and habitat degradation. Three major roads and a railway track pass through the Gir Protected Area, the Asiatic Lion's last remaining habitat.

Also, there are three big temples inside the PA that attract large number of pilgrims, particularly during certain times of the year.

But notably, the steps taken by the PA management and the Government of Gujarat, have led to an overall improvement in the habitat conditions and the population of lions has been increasing steadily since 1974.

One such step was the relocation of about 50 maldhari nesses outside Gir, which led to improvement in lion population and its habitat.

But the increased population has resulted in their spill over the PA. Therefore, at present, the most pressing threat to the lion population of the Gir PA comes from the increasing hostility towards them from the resident human population. Due to the increase in population, about 100 lions stay outside the PA and face conflicts with humans. Though the conflict is not high now, with changing lifestyles and values these can increase in future.

An emerging threat is the number of lions falling in the open wells in the fields around Gir NP. The main reason is that wells in arable fields are unguarded. These wells have been made at ground level without any protection like parapet walls around them.

Ultimately, this single population is very susceptible to catastrophic events. Increasing the effective range of lions by connecting protected areas with lion-friendly corridors and establishing a second lion population elsewhere in India through translocation are required management options that need to materialize in the medium term future.

Despite the tiny population and limited range of the Asian lion, there are actually some positive signs that conservation attempts are paying off. The current population of about 411 individuals is twice what the total population was three decades ago.

The Asiatic Lion Reintroduction Project is attempting to restore additional wild populations with captive-bred animals, though the state of Gujarat is resisting this as they wish to retain their unique status as the only home of wild Asian Lions.

Conservation

Conservation efforts for Asiatic Lion were made for the first time in the year 1910. The Nawab of Junagadh imposed a ban on the hunting of lions within the boundaries of his province. The ban was continued even when India gained independence in 1944. In 1960s and 1970s, Gir forest, the home of the last surviving Indian lions, was converted into a National Park and Sanctuary. Presently, Kuno Project is being undertaken with a view to reduce the overcrowding at Gir.

The park boundaries can be consolidated to prevent the human-lion clash. Much attention needs to be paid on the protection of the Asiatic lions from poachers. It is becoming one of the greatest threats to the life of the lions. The natural habitat of Gir National Park also needs to be protected. The afforestation programmes that were earlier carried out in the grasslands and savannah areas of Gir need to be avoided as far as possible. Efforts should be made to open up forest canopies and to increase the range of the Gir PA (Protected Area).

Establishment of at least one other wild population is advisable for population safety, for maximizing genetic diversity, and in terms of ecology (re-establishing the lion as a component of the fauna in its former range). However, there are problems in attempting this: a previous attempt to establish a second subpopulation in the Chandraprabha Wildlife Sanctuary in eastern Uttar Pradesh appeared to be succeeding, as the population grew from three to 11 animals, but then the lions disappeared, presumably shot or poisoned (Nowell and Jackson 1996).

Palpur-Kuno Wildlife Sanctuary in northern Madhya Pradesh has now been selected as the best candidate area. Communities will require resettlement to make room for the Lions, but this time great care is being taken to make the process participatory and to attempt to satisfy local needs, and not engender hostility towards Lion conservation. It has become necessary to involve people in the conservation efforts being planned for the Indian lions. A step that needs to be taken in this regard is to include the local residents within the field staff of the Gir Protected Area.

It will not only sensitize people about the majestic species, but also educate them about the need to protect the species from extinction. At the same time, efforts should be undertaken to relocate as many humans out of the protected area of Gir as possible. Included on CITES Appendix I. This subspecies is fully protected in India (Nowell and Jackson 1996).

The need for conservation of wildlife in India is often questioned because of the apparently incorrect priority in the face of direct poverty of the people. However, Article 48 of the Constitution of India specifies that, "The state shall endeavour to protect and improve the environment and to safeguard the forests and wildlife of the country" and Article 51-A states that "it shall be the duty of every citizen of India to protect and improve the natural environment including forests, lakes, rivers, and wildlife and to have compassion for living creatures

Reintroduction

The **Asiatic Lion Reintroduction Project** is an initiative of the Indian Government to provide safeguards to the Asiatic lion (*Panthera leo persica*) from extinction in the wild by means of reintroduction. The last wild population of the Asiatic Lion is found in the Gir Forest region of the state of Gujarat. The single population faces the threats of epidemics, natural disasters and other anthropogenic factors. The project aims to establish a second independent population of Asiatic Lions at the Kuno Wildlife Sanctuary in the Indian state of Madhya Pradesh. However, the proposed translocation is unpopular in Gujarat and has been bitterly contested by the state government.

The distribution of Asiatic Lion, once found widely in West and South Asia, dwindled to a single population in the Gir Forest National Park and Wildlife Sanctuary in India. The population at Gir declined to 18 animals in 1893 but increased due to protection and conservation efforts to 284 in 1994. The Gir Wildlife Sanctuary is now highly overpopulated with lions, the 2010 census showed the strength to be 411 lions.

There are numerous deaths in the population annually because of ever increasing competition between the human and animal overcrowding. Asiatic lion prides require large territories but there is limited space at Gir wildlife sanctuary, which is boxed in on all sides by heavy human habitation. Gir lions have started moving outwards from the sanctuary and establishing homes outside the protected areas.

The lions are now spread over 16,000 square kilometres (6,200 sq mi) in the vicinity of 1050 villages in three contiguous districts – Amreli, Bhavnagar and Junagadh.

The sites were assessed and ranked for suitability as follows:

- ☆ Kuno Wildlife Sanctuary – found most suitable for reintroduction.
- ☆ Sita Mata Wildlife Sanctuary – later rejected due to human interference and inadequate prey population.
- ☆ Darrah - Jawahar Sagar Wildlife Sanctuary – later rejected due to degraded habitat and unsuitable geography.
- ☆ Kumbhalgarh Wildlife Sanctuary – assessed as having limited area, unsuitable terrain, limited water and prey base as well as disturbance.
- ☆ Barda Wildlife Sanctuary – assessed as having scarcity of water, prey and forage, as well as encroachment and disturbance

Translocation of Asiatic Lions

Wildlife is under threat in most of India, but there's one state that's clinging to its fauna, if rather too tightly. The state of Gujarat – whose Gir forest shelters the world's only Asiatic lion population – has lost a bitter battle over an ambitious translocation project.

For over 18 years, conservationists have been attempting to move a pride of Gir lions to the Kuno sanctuary in the state of Madhya Pradesh. But the Gujarat government stubbornly refused to let the lions go. Meanwhile, an impatient Madhya Pradesh government bizarrely suggested introducing African cheetahs, whose Asiatic cousin once roamed the area.

Finally, on 15 April, after eight years of litigation, the Supreme Court of India cleared the translocation of the lions, though it said an emphatic no to the cheetahs. Now the reluctant Gujarat government is considering filing a review petition, even as angry locals protest the translocation, some even threatening to kill themselves.

The Gujarat government, and Chief Minister Narendra Modi, have been arguing for years that the Gir lions are the "pride of the state". In court, the lions were described as "family members who could not be parted with." The native Maldhari tribals have lived alongside the lions for years, on the whole peacefully. Even the Supreme Court acknowledged the state's role in saving the lion from an extinction that seemed almost certain only a few decades ago. The latest census put the lion population at 411, up from 359 in 2005.

But most wildlife experts agree that while the state has done well, the lions have outgrown the Gir forest, and a single epidemic or calamity could wipe out the entire population. In 1994, an epidemic of canine distemper killed an estimated 1000 lions in the Serengeti National Park in Tanzania.

"The Gir conservation project has staved off extinction and helped increase population. The translocation is about strengthening conservation prospects and managing success. At the moment, all our eggs are in one basket and that is a huge risk," says Dr Ravi Chellam, well known conservationist and former director of the Wildlife Conservation Society of India. Chellam was one of the experts who recommended the translocation, which he likens to buying life insurance.

In its judgement, not surprisingly, the Supreme Court rejected the Gujarat government's emotional appeal. "The cardinal issue is not whether the Asian lion is a family member, but the preservation of an endangered species," ruled the court, while directing that the lions be moved within six months.

Meanwhile, now that the cheetahs have been nixed, the Gujarat government reportedly plans to argue that Kuno is already home to stray tigers from the nearby tiger sanctuary of Ranthambhore. Will two big cats in the same area cause trouble? Most experts say the tiger population in Kuno is too small to make a difference. "Tigers and lions have had overlapping distribution historically.

It's still not a good enough reason to deny translocation," says Chellam.

An earlier translocation of lions to the state of Uttar Pradesh in the fifties failed. With insufficient prey and only a small area to roam, the lions hunted nearby village cattle, and were hunted in their turn by angry villagers. This time around, conservationists say things will be different. "Kuno was chosen because of its size–3000 sq km-and diverse prey base. Lions need lots of space, plenty of prey, and protection from people," says Chellam. Wildlife studies have shown that the prey count in Kuno is actually higher than that of Gir. 24 villages were relocated in Kuno, at considerable expense, to make way for the lions, though many villagers complain that they have still not received compensation.

Despite the years of preparation, the hard work has only just begun. Indeed, Kuno is a test case for conservation through translocation. Kuno has a strong gun culture, and the Supreme Court has agreed that forest staff will need to be trained and deployed to prevent poaching. Locals will need to be educated, involved and given a stake in conservation, just as the Maldharis had. The project will need male lions to be moved from Gir to Kuno every three to five years for the next 30 years.

All this means that Gujarat will very soon have to stop sulking and start cooperating. After nearly 20 years of bickering and debate, it's time to put everything else aside and act to save the Asiatic lion.

Pride in Place: Reintroduction of Asiatic Lions in Gir Forest, India

Mark Paulson

The Gir Protected Area - Gujarat, India

Asiatic lions (*Panthera leo persica*) once ranged from Syria through India and Iran to as far west as Greece until the middle of the 19th century, when firearms became universally widespread (Joslin, 1985). Hunting and habitat loss have since compromised the species' populations to the extent that by the turn of the century they were found only in a relatively small portion of western India. The population descended to its lowest number in 1893 with 18 remaining individuals (Saberwal *et al.*, 1994). Protection efforts have since raised that number from 177 in 1968 (Joslin 1985) to 284 in 1994 (Chellam and Johnsingh, 1994). Asiatic lions now dwell almost exclusively in the Gir forest of northwestern India, located in the Junagadh district of Saurashtra, Gujarat State (Saberwal *et al.*, 1994). The Gir forest has been reduced to almost one- third its original size and now covers about 1400 km^2 of semiarid country. Contained within the Gir forest are the Gir Wildlife Sanctuary (established in 1965), and Gir National Park (established in 1975), which together make up the Gir Protected Area or "PA". The area is of high conservation value as the last remaining example of native landscape within the Saurashtra peninsula (Chellam and Johnsingh, 1994). Restoration efforts for Asiatic lions are complicated by their current low numbers and habitat availability restrictions. Current plans are directed towards managing lion-human conflicts and restoring water hole habitat. Future restoration attempts must also advance lion translocation efforts and integrate Maldhari involvement in the management of the species.

The first significant effort toward protection for the Asiatic lion occurred in 1900 when the Nawab of Junagadh banned lion hunting on his private landholdings

(Meegama 1979). After Indian independence in 1947 the area was declared a preserve and achieved full sanctuary status in 1965. Numerous studies have been done on the Gir lions and their habitat, but the essential conservation issue was first addressed in the early 1970's when lion predation studies began to focus directly on human factors regarding lion-livestock depredation and lion-on-human attacks.

A 1973 study found that 75 per cent of the lion's diet was livestock while wild ungulates made up only 25 per cent (Joslin 1985). About 2,200 local grazers, the Maldharis, have long endured losses in livestock and in human life, while many of their grazing practices and traditional customs have contributed to lions-human altercations. Stabilization of the Asiatic lion population and the reclamation of its dwindling habitat has thereafter focused on maintaining a workable human-lion coexistence.

WWF-India Interventions

The Asiatic lion is a priority species for conservation for WWF-India. Project has been initiated to construct barricades around open wells around the Gir NP to decrease the incidences of lions falling into such wells. The aim is towards long-term conservation of the species, and this also includes having a viable wild population of Asiatic lion at an additional place.

The following aspects are important for long term conservation of Asiatic lion in India:

1. Strengthening overall protection measures
2. Habitat management in Gir
3. Mitigating Conflicts outside Gir PA
4. Greater Gir concept
5. Relocation of villages Jambuthala, Timbarwa and Ghodavadi villages.

New Plans

Lion attacks on humans following the drought forced the restoration issue for lion conservationists. Locals, lacking technical ecological criteria to validate lion behavior, were forced into direct conflict with both lions and GFD officials, increasing tensions on all sides.

More aggressive strategies would be required. The new plan proposes (Chellam and Johnsingh, 1994):

- ☆ Immediate and consistent response to all human based lion problems. Poachers, illegal harvesting of firewood or grazing within the park, etc.
- ☆ Re-evaluation of Maldhari relocation plans. Emphasis on decreasing edge effects by reducing the number of private landholdings that pierce the interior of the PA should be sought. If officials decide to continue with the relocation plan, the gradual pace must be maintained to soften the effect of livestock disappearance from lion prey options.

- Habitat restoration at water hole locations and riverine forest sites should become a high priority. Native teak (Calophyllum elatum) harvesting has been proposed as an option toward habitat reclamation (Gadgil 1992). Some dispute remains over whether lions are inhibited in the hunt in dense underbrush. Pruning of shrubbery around watering holes to improve hunting habitat has not been wholly embraced.
- Maldari involvement in the GFD should be encouraged as well as their involvement in new ecotourism options. While some care must be taken in incorporating ecotourism into traditional lifestyles, limited use of the methods may succeed if Maldharis become beneficiaries from that source. Existing attempts at educational tourism require more funding and adequate resources as well as a working site that is closer than the current location west of the park at Sasan.
- Saberwal (1994) recognizes the need for improving public opinion regarding living with lions and suggests that financial incentive should be offered "by the government to all residents of subdistricts that abut the Gir forest, possibly in proportion to lion densites." These management alterations have been proposed as improvements to existing habitat. Most biologists involved in Asiatic lion research also support lion translocation in an attempt to meet a larger goal of population stability.

Conclusion

Further steps towards restoration of Panthera leo persicaand its habitat need to address some key biological needs. Existing habitat should continue to be managed for restoration of native plant species that support chital and other native lion prey, especially around water holes and other favoredhunting spots. Purchasing private land holdings that intrude into the PA would be a great advancement towards reducing edge effects and lion-human interactions, although it will likely be an expensive one. Lion translocation seems to be the most immediate long range need. Hopefully one of the many study locations for establishing new prides will prove viable.

Redirecting efforts in solving human-lion interactions seems critical. Maldharis have in the past shown a willingness to risk some lion hazards to human life and livestock losses. If the number of interactions could be reduced, a more sustainable lion-human relationship could be reclaimed.

Efforts on this behalf should probably be directed toward translocation and Maldhari relocation initially. Included in any management plan should be immediate economic compensation for established losses and aggressive attempts to employ Maldharis in management, either through the GFD or a newly established bureau. These tactics may serve to improve the local opinion of lions in the form of reducing economic losses and in lowering attacks on people.

It seems likely that only when Maldharis perceive the protection of Gir lions as a public good will a long-range, effective preservation programmes will be successful. As Kellert (1985) points out:

"Most species are endangered not because of their biological inadequacies, but because of a variety of human social, psychological, and cultural factors". Recovery plans should integrate lion habitat and population management with the difficult objectives of socio-economic development for the Maldhari

References

Ashraf, N.V.K., Chellam, Ravi, Sharma, D., Molur, Sanjay, Walker, Sally, eds. (May 1995). "Asiatic Lion Report". *Population and Habitat Viability Assessment P.H.V.A. and Global Animal Survival Plan Workshops (18–21 October 1993, Baroda, India)*. Zoo Outreach Organisation/CBSG, India. *Retrieved* 25 April 2013.

Chellam, R. and A.J.T. Johnsingh (1993). Management of Asiatic lions in the Gir Forest, India. Symposia of The Zoological Society of London 65: 409–424. Gadgil, M. (1992).Conserving biodiversity as if people matter: : a case study from India. *Ambio* 21(3): 266–270.

Chellam, Ravi, Johnsingh, A.J.T. (1999). "Translocating Asiatic Lions, India". *Re-introduction News* (18): 11–13. ISSN 1560–3709. *Retrieved* 19 April 2013.

Johnsingh, AJT (1 February 2006). *Field Days: A Naturalist's Journey Through South and Southeast Asia*. Universities Press. pp. 126–138. ISBN 978–81–7371–552–5. *Retrieved* 22 April 2013.

Joslin, P. (1985). The environmental limitations and future of the Asiatic Lion. *Journal of The Bombay Natural History Society* 81(3): 648–664.

Kellert, S.R. (1985). Social and perceptual factors in endangered species management. Journal of Wildlife Management 49(2): 528–536. Khan, J.A. (1995). Conservation and management of Gir Lion Sanctuary and National Park. *Biological Conservation* 73(3): 183–188.

Mazoomdar, Jay (27 April 2013). "Science Trumps Politics". *Tehelka magazine*. Tehelka.com. *Retrieved* 19 April 2013.

Meegama, N.C. (1979). The Asiatic lion – the last refuge. Loris 15(1): 35–36

Parmar, Vijaysinh (15 July 2013). "Asiatic lions have homes in over 1,000 Saurashtra villages". *ndiatimes.com* (Bennett and Coleman Ltd). *Retrieved* 23 July 2013.

Paulson, Mark (1999). "Pride in Place: Reintroduction of Asiatic Lions in Gir Forest, India". *Restoration and Reclamation Review* (Dept of Horticultural Science, University of Minnesota) **4** (3). *Retrieved* 19 April 2013.

Shah, Jumana (16 April 2013). "Asiatic lions: Gujarat's roar in court lacked substance". *DailyBhaskar.com*. DB Corp Ltd, Noida, UP, India. *Retrieved* 19 April 2013.

Chapter 21

Running Out of Time: The Decline of Great Indian Bustard

Aeshna Nigam and Ankita Dua

Department of Zoology,
Shivaji College, University of Delhi, New Delhi

"If we kill off the wild, then we are killing a part of our souls"

— ***Jane Goodall***

ABSTRACT

The great Indian Bustard, one of the largest species of birds is endemic to the Indian subcontinent. The last fifty years have seen a drastic decline in their population till in 2011 the IUCN declared it a critically endangered species. The present global population of great Indian Bustard is approximately 150 of which more than 70 per cent are inhabitants of Rajasthan. Hunting, poaching and habitat destruction are the major causes of the steep decline in their population. The government has launched many initiatives to save the bustards, the major being 'Project Great Indian Bustard' by state government of Rajasthan. There need to be more accelarated efforts in carrying out conservation breeding programmes or these tall handsome birds will soon become extinct.

Keywords: *Great Indian Bustard, India, Pakistan, Conservation, Habitat, Critically endangered.*

Introduction

Characteristics of the Great Indian Bustard

Habitat:	Short grass plains and deserts in dry arid area
Appearance:	One of the large birds in pale colour with a black cap
Size and Height:	Head-body length: 92–122 cm (1m)
Weight:	Ranges between 3.5–15 kg (males: 8–15 kg, females: 3.5–7.75 kg)

Wingspan :	2.1–2.4 m (male)
Coat:	Male is usually sandy buff colored and develops black breast band during breeding season. Females are dull in color with greyer neck and absence of breast band.
Litter size:	1–3 eggs. Sex ratio at birth roughly 1.5:1 (female: male).

Runner up in the race to become the National bird of India in 1969, Great Indian Bustard (GIB) today faces the danger of extinction due to anthropogenic activities (Ali, 1961). This tall, handsome bird of the family Otididae is an obligate grassland inhabitant with open nesting system and exhibits k-selected traits (Lack, 1954). These characteristics render them vulnerable to extinction when subject to environmental threats and human activities. The scientific name of GIB is *Ardeotis nigriceps* and is popularly known as Maldhok, Yerbhoot, Ghorad, Tuqdar, Sohan chidia, Sone chiraya etc. It is state bird of Rajasthan where it is known as Godawan.

GIB is endemic to the Indian subcontinent with their population further fragmented into 6 states of India (Rajasthan, Maharashtra, Andhra Pradesh, Gujarat, Karnataka and Madhya Pradesh) and Pakistan. Their population was roughly estimated at 1260 in 1969 (Dharmakumarsinjhi, 1971) and drastically reduced in the span of 10 years to 745 in 1978 (Dharmakumarsinjhi, 1978a). In the beginning of the 21st century the numbers dwindled to 600 (BirdLife International 2001), 300 in 2008 and the current population is estimateed be 150 (Birdlife International 2013). The bulk of the population is found in Rajasthan with scanty distribution of less than 35 birds in the other states.

The GIB is one of the heaviest flying birds. The origin of bustards took place 77 million years ago when they inhabited the old world grasslands (Johnsgard, 1991). Later they coevolved with the wild ungulates and depend on the grazers to maintain their habitat structure. The GIB habitats are wide sparse grass-scrub landscapes with minimal agricultural activities. Over the last thousand years, the domestication of animals and encroachment of land for agricultural purposes has resulted in the incidences of overgrazing and destruction of their habitat (Skarpe 1991, Dutta 2010).

The GIB exhibit significant sexual dimorphism (Figure 21.1). The males are 122 cm long standing upto a height of 1 m and weigh approximately 18 kg. Their sandy brown and white colored feathers with long white neck, grey head and long bare legs give them ostrich-like appearance. The males have a black breastband and white belly. The females are about one third the size of males with length of 92 cm and weigh about 7 kg. They have a duller appearance with slender neck and breastband either rudimentary or broken (Jerdon, 1864). This bird flies low with slow heavy sweeps of wings and never sails (Finn, 1916). They have excellent hearing ability and eyesight which helps them to locate their prey and hide from danger. Male and female birds move in separate unisexual flock which varies from 3–10 birds. During the breeding season, the cock acquires a harem of 3–4 females. It struts to attract the females and expand the gular pouch to produce resonant mating call which can be heard upto a distance of 50 0m (Figure 21. 2). They breed and make their nest on open grounds. The female lays 1–3 eggs per year with an average

Male Great Indian Bustard

Female Great Indian Bustard

Figure 21.1: Distinct Sexual Dimorphism Exhibited by the Two Genders of the Great Indian Bustard. The first picture is of the male with sandy buff colour and breast band. The female is comparatively short and dull in colour and lacks breast band. Photo credits: Joanna Van Gruisen (male) and Sumit Sen (female).

incubation period of approximately 27 days (Rahmani, 1986). The male does not participate in incubation of eggs.

The bustards are oppurtunistic omnivores. Their food majorly comprises of grass, seeds, pods of legumes, beetles, berries, grasshoppers and groundnuts but changes with the availability of seasonal food. Sometimes they also feast on small

Figure 21.2: Elaborate Courtship Ritual by Male. The first picture shows inflation of gular pouch for resonating mating call. The second picture shows the female inspecting the male for mating. Photo credits: Mohanram Kemparaju and Franz Cova.

rodents, insects and reptiles. Like any other desert inahabitant, GIB is a facultative drinker and drinks regularly when water is available but can survive long dry spells.

The GIB is considered as an indicator of the health of grasslands which often remain neglected and are considered wastelands. However, these grasslands are major requirement of the grazing livestock and hence reflect on the economy of a country. Historically, the GIBs have been subject to game hunting and poaching which caused a decline in their population. This process has been accelarated in modern times with vehicular access to their undisturbed habitat, encroachment of land for agricultural practices, livestock overgrazing and collision with power lines. The small clutch size, slow breeding habits and > 40 per cent mortality rate of young ones during first year makes it even more difficult to save their declining population. Thus it is important to design conservation strategies keeping all these factors in mind to save GIB and its sister taxa – MacQueen's bustard, Lesser Florican and Bengal Florican (other three species of bustards in India which have been labelled critically endangered and threatened) or else they will become extinct in India like cheetah.

Battle for Survival

The critically endangered status of the GIB is owed to the alarming dip in the numbers of the species as a result of habitat loss, degradation and hunting activities. The GIB is included in Schedule I of the Wildlife (Protection) Act of India. At the World Conservation Congress in Thailand, in 2004, the IUCN had passed a resolution to urge and take immediate action towards conservation of bustards.

The most important causes that are threatening the existence of the game bird are enumerated as follows:

- ☆ Habitat loss: lack of maintenance of arid grasslands. More than 90 per cent of the geographic range of the species has been destroyed as a result of urbanization and development.
- ☆ Agricultural expansion and large scale farming techniques.
- ☆ Development of infrastructure in areas that were originally the niche for these birds.
- ☆ Mining and industrialization
- ☆ Drastic errors in habitat management; for example: the situation arising in and around the Desert National Park. A shift in crop pattern has been initiated that is not bustard friendly! Traditional monsoon crops have been shifted to cash crops, which aren't suitable for the bird population.
- ☆ Poaching is a prevalant cause of decline of populations in India as well as Pakistan.
- ☆ Natural predators such as Egyptian vultures, jungle cats, foxes, crows and mongoose are a big threat to the nesting females (Ghouse and Indira, 2015). Presence of blackbucks increases the chances of egg trampling as well.

In order to assess the present genetic diversity of this game bird, a number of techniques have been employed. A detailed study of the hypervariable regions of the mitochondrial DNA of samples collected from five states within the geographic distribution range of the bird in India was done to investigate the current diversity (Ishtiaq *et al.*, 2011). Alarming results of low genetic diversity led to the hypothesis of a bottleneck event that has resulted in this immense loss of species. The bird has made it to the critically endangered category in 2011, post the estimation of its declining numbers by Bird Life International and IUCN. Lower genetic diversity is a direct indication that the species is more vulnerable to ecological threats and may not be able to adapt itself to changing ecological conditions.

Conservation Strategies

An estimated 50-150 mature birds still remain in the country and mostly limited to the state of Rajasthan. Looking at the precarious numbers, the Indian government launched major initiatives to save the bustard along with its relatives Bengal florican (*Houbaropsis bengalensis*) and the lesser florican (*Sypheotides indicus*). Their habitats all over the country are shown in Figure 21.3.

The state government of Rajasthan on 5th of June, 2013 launched a 12 crore project to make a sincere effort to save its state bird (Figure 21.4). "Project Great Indian Bustard" aims to work towards the conservation of this neglected species that is more vulnerable to extinction than the tiger, Asiatic lion and Florida panther.

Along with Rajasthan, around 30-odd GIBs survive in the Abdasa taluka of Kutch district. Here, the Kutch Bustard Sanctuary is spread over a total of 2 sq km. The areas surrounding the sanctuary are for breeding and display activities.

Efforts to Tackle the Situation

The Ministry of Environment and Forests, Government of India has initiated an effort of setting up Bustard Conservation Committees (BCCs) across states of the country and declare potential breeding zones of the bustard populations as "Critical Bustard Areas".

- Restricting anthropogenic activities in spheres where the habitat is apt for the bird to survive. Making specific enclosures at such places that will protect the bird from poachers as well. These must be fenced and daily monitored. Involving local people in conservation strategies and management is essential as they can help in these tasks.
- Enriching the habitats by planting crops, grasses that favor the survival and expansion of the species.
- Rehabilitation of grasslands; any invasive species of tall shrubs/trees must be avoided in their proposed habitat that may cause hinder the visibility of both the sexes during mating. Also overgrazing of these grasslands by cattle must be avoided.
- An effort to do captive breeding in a scientific manner can aid conservation of this species. Breeding programmes must be designed by competent

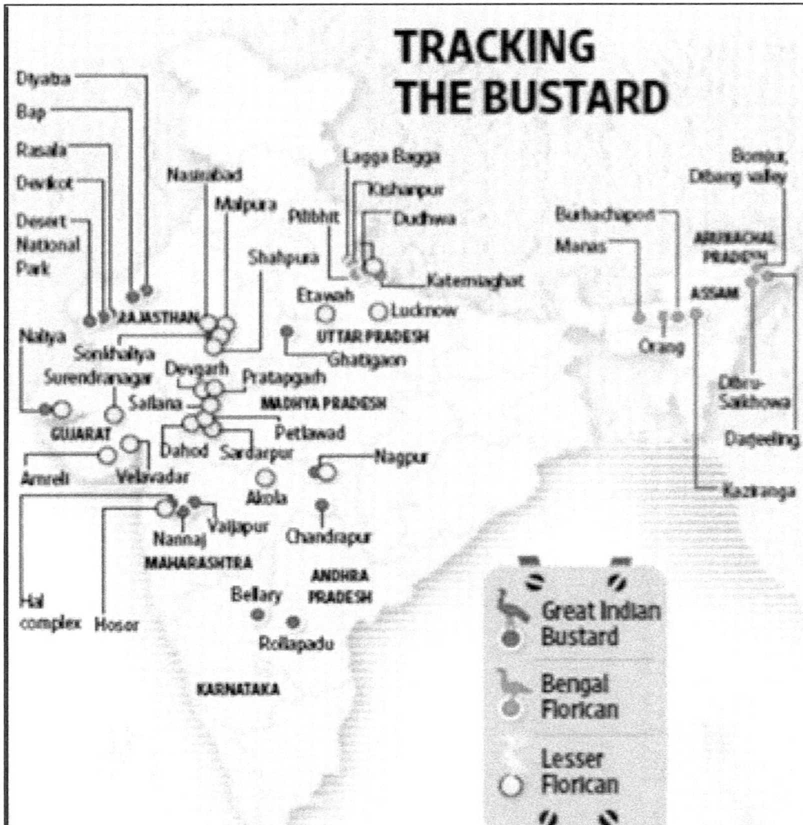

Picture 3: a. Tracking the locations of Great Indian Bustard, Bengal Florican & Lesser Florican over the country. Source: Bombay Natural History Society

b. Declining numbers of Great Indian Bustard over the past few decades.

Figure 21.3: a: Tracking the locations of Great Indian Bustard, Bengal Florican and Lesser Florican over the country (Source: Bombay Natural History Society); b: Declining Number of Great Indian Bustard over the Past Few Decades.

Figure 21.4: Project Great Indian Bustard initiated by the State Government of Rajasthan.

authorities and the personnel that are a part of it must have undergone training specializing in *ex situ* strategies.

- ✰ Bustard gene banks can also be set up as secondary programmes.
- ✰ Core breeding areas, feeding sites, lek sites need to be identified and any kind of human interference here during the breeding months (July-September) needs to be banned. Entry of tourists or photographers must be banned during mating season.
- ✰ Infrastructure that harms the bird population should be not allowed within a 2 km radius of the breeding areas that have been identified. Strict legislations must monitor active implementation of these rules.
- ✰ It is essential to have various awareness programmes to educate the local populations about the graveness of the matter. Groups must be formed at the community level that will help in this effort. Also, alternative sources of livelihood may be encouraged to prevent any kind of poaching of these species.

Conclusion

Efforts are being made substantially in the direction of formulating *in situ* and *ex situ* conservation strategies. The government has finally woken up to the plight of the game bird and hopefully strict measures and legislations will pave the way for saving the GIB population. One great such effort has been displayed by Pune based ornithologist who raised £52,000 for the conservation of GIB. He has been awarded the Whitley Award also known as the "Green Oscar".

Even though the GIB is on the brink of extinction by the passing minute, all is not lost if sincere steps are taken. Strategic conservation strategies have saved species left in even lower numbers, such as the California Condor, Mauritius parakeet and the Norfolk island green parrot. Hence, the chances are bleak but all is not lost as initiatives like the "Project Great Indian Bustard" have come up in this direction. Hopefully all these endevours will reap fruit and one of our country's biggest conservation challenges will be won.

References

Ali S (1961). Our National Bird. Newsletter for Birdwatchers 1(4): 3–4.

BirdLife International (2001). Threatened birds of Asia: the BirdLife International red data book. BirdLife International, Cambridge.

BirdLife International (2013). "*Ardeotis nigriceps*". IUCN Red List of Threatened Species. International Union for Conservation of Nature 2: *Retrieved* 26 November 2013.

Dharmakumarsinjhi RS (1971). Study of general great Indian bustard. Final report. WWF. Morges.

Dharmakumarsinjhi RS (1978a). Report on the great Indian bustard. ICBP bustard group unpublished circular. Birdlife International, Cambridge

Dutta S, Rahmani AS and Jhala, YV (2010). Running out of time? The great Indian bustard *Ardeotis nigriceps*–status, viability and conservation strategies. *European Journal of Wildlife Resarch* 57(3): 615–625

Finn F (1916). Game Birds of India and Asia. Calcutta: Thacker, Spink and Co.

Ghouse SM, Indira P (2015) Conservation Strategies of Great Indian Bustard–Ardeotis nigriceps of Rollapadu wildlife sanctuary in India: An Overview. *International Journal of Pharmaceutics and Drug Analysis* 3 (2): 27–33.

Guidelines for preparation for state action plan for bustards' recovery programme. Government of India, Ministry of Environment and Forests.

http://timesofindia.indiatimes.com/city/pune/Pune–ornithologist–raises–52000–to–save–the–Great–Indian–Bustard/articleshow/47335145.cms

http://www.birdlife.org/datazone/speciesfactsheet.php?id=2767

http://www.britannica.com/EBchecked/topic/243369/great–Indian–bustard

http://www.deccanherald.com/content/458761/indian–bustards–facing–extinction.html

http://www.rajforest.nic.in/?q=project–great–indian–bustard

http://www.sanctuaryasia.com/magazines/features/9260–critically–endangered–why–the–great–indian–bustard–might–need–a–captive–breeding–programme

http://www.sciencelog.net/2015/02/great–indian–bustard–in–india.html

http://www.wwfindia.org/about_wwf/priority_species/threatened_species/great_indian_bustard/

Ishtiaq F, Dutta S, Yumnam B, Jhala YV (2011). Low genetic diversity in the endangered great Indian bustard (*Ardeotis nigriceps*) across India and implications for conservation. *Conservation Genetics* 12 (3): 857–863.

Jerdon TC (1864). The birds of India. George Wymann and Co: 607–611.

Johnsgard PA (1991). Bustards, Hemipodes, and Sandgrouse. Birds of Dry Places. Oxford University Press, New York, USA.

Lack D (1954). The natural regulation of animal populations. Oxford University Press, London.

Rahmani AS and Manakadan R (1986). Movement and Flock composition of the Great Indian Bustard *Ardeotis nigriceps* (Vigors) at Nanaj, Solapur District, Maharahtra, India. *Journal of Bombay Natural History Society* 83: 17–31.

Skarpe C (1991). Impact of grazing in Savanna ecosystems. *Ambio* 208: 351–356.

Index

C

D

E

S

T

U

V

W

www.ingramcontent.com/pod-product-compliance
Ingram Content Group UK Ltd.
Pitfield, Milton Keynes, MK11 3LW, UK
UKHW020250290926
15470UKWH00002B/854

9 789386 071330